SPYING with MAPS

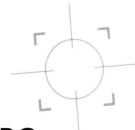

SPYING with MAPS

SURVEILLANCE TECHNOLOGIES AND
THE FUTURE OF PRIVACY

Mark Monmonier

THE UNIVERSITY OF CHICAGO PRESS
CHICAGO AND LONDON

The University of Chicago Press, Chicago 60637
The University of Chicago Press, Ltd., London
© 2002 by The University of Chicago
All rights reserved. Published 2002
Paperback edition 2004
Printed in the United States of America

11 10 09 08 07 06 05 04   2 3 4 5

ISBN: 0-226-53427-8 (cloth)
ISBN: 0-226-53428-6 (paperback)

Library of Congress Cataloging-in-Publication Data

Monmonier, Mark S.
    Spying with maps : surveillance technologies and the future of privacy
    / Mark Monmonier.
        p. cm.
    Includes bibliographical references and index.
    ISBN 0-226-53427-8 (cloth : alk. paper)
    1. Electronic surveillance.   2. Remote sensing.   3. Privacy, Right of.
    I. Title.
    TK7882.E2 M65 2002
    621.389′28—dc21

                                                        2002018124

⊗ The paper used in this publication meets the minimum requirements of
the American National Standard for Information Sciences—Permanence of
Paper for Printed Library Materials, ANSI Z39.48-1992.

For Will and Ruby Miller, geographers extraordinaire

Contents

Acknowledgments

In writing this book, I benefited from the insights of Margot Ackley and Doug van de Kamp, NOAA Forecast Systems Laboratory, Boulder, Colorado; Harry Carlson, Department of Public Works, Syracuse, New York; Jerry Dobson, Oak Ridge National Laboratory, Oak Ridge, Tennessee; Gary Hufford, National Weather Service, Alaska Region, Anchorage, Alaska; David Miller, Texas Department of Transportation, Amarillo District; Paul G. Richards, Mellon Professor of Natural Science, Lamont-Doherty Earth Observatory, Columbia University; and Paul L. Robinson, Oklahoma State Bureau of Narcotics and Dangerous Drugs Control. No less useful was the enthusiastic probing of the students in my fall 2000 seminar: Christian Axsiom, Anna Dolmatch, Tracy Edwards, Lillian Jeng, Hilary McLeod, Rich van Deusen, and Tom Whitfield. For illustrations or data I am also grateful to Nancy Adams, Eastman Kodak Company;

Andrew Etkind, Garmin International; Keith Harries, Department of Geography, University of Maryland, Baltimore County; Jolene Hernon, National Institute of Justice; Amy L. King, Geographic Information Science and Technology Program, Oak Ridge National Laboratory; Eric Lund, Veris Technologies, Salinas, Kansas; Lloyd Novick, Commissioner of Health, Onondaga County, New York; and John R. Schott, Remote Sensing and Imaging Laboratory, Rochester Institute of Technology. A one-semester research leave from Syracuse University's Maxwell School of Citizenship and Public Affairs provided valuable time to read, think, and ask questions. At Syracuse University, Becky Carlson and Joe Stoll assisted with scanning and fonts.

My longstanding relationship with the University of Chicago Press is a priceless asset. Penny Kaiserlian, my longtime editor and now director of the University Press of Virginia, was an early backer of the project, and Christie Henry, my current editor, offered encouragement and insightful suggestions. External readers Marc Armstrong, Harlan Onsrud, and Nancy Obermeyer made helpful comments on the penultimate draft; Jenni Fry dealt diligently with ephemeral URLs and other glitches in the manuscript; and Renate Gokl and Russell Harper guided the book through production. I also value the continuing support of Alice Bennett, Mike Brehm, Paula Duffy, Erin Hogan, and Carol Kasper. And at Waldorf Parkway, there's Marge, whose patience I've yet to exhaust. As country musicians revel in saying: Thank you, all.

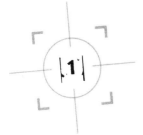

Maps That Watch

Privacy and *mapping* are two words that rarely share the same sentence. After all, what do most of us have to hide that anyone would want to map? But toward the end of the last century, cartographers added privacy to a growing list of policy issues that includes copyright, liability, and public access. Mapping, it turns out, can reveal quite a bit about what we do and who we are. I say *mapping*, rather than *maps*, because cartography is not limited to static maps printed on paper or displayed on computer screens. In the new cartographies of surveillance, the maps one looks at are less important than the spatial data systems that store and integrate facts about where we live and work. Location is a powerful key for relating disparate databanks and unearthing information about possessions, spending habits, and an assortment of behaviors and preferences, real or imagined. What's more, these electronic maps are becoming increasingly detailed and timely, if not more reliable. What gets into the system as well as who can use the data and for what purposes makes privacy in mapping a key concern of anyone who fills out surveys, owns a home, or registers a car or firearm.

I could write this book to frighten readers, but I won't. However odious the threat of rampant snooping or a new holocaust, fear founded on mere possibility is less helpful than wariness grounded in understanding. Informed skepticism about cartographic surveillance should encourage the vigorous yet vigilant application of this ambiguous technology that, like the bulldozer and the chemical plant, can—if controlled—do far more good than harm. If this ambiguity is disconcerting, get used to it. A jeremiad that capitulates to gloom and doom would be no better than an equally naïve celebration of trouble-free progress.

A Luddite rant would also ignore some fascinating stories of fortuitous discoveries and unintended consequences. As the following chapters reveal, there are multiple cartographies of surveillance, some concerned primarily with integrating databases, some involving satellite imagery or satellite-based location tracking, and some narrowly focused on specific needs, like growing crops or controlling crime. Although all applications examined here use monitoring to control human behavior—that's the definition of *surveillance* —the behaviors in question range from the predations of war and crime to economic decisions about when to plant and where to spread fertilizer. Big Brother is doing most of the watching, at least for now, but corporations, local governments, and other Little Brothers are quickly getting involved.

If you don't see the danger, think integration. The threat to personal privacy lies mainly in the imminent ease of linking a large number of databases rapidly and reliably in order to track shipments, pollutants, lost children, potential terrorists, campaign contributions, or anyone walking around with a cell phone. An invasive system would not only monitor location in real time but also store the data indefinitely to reconstruct an individual's movements during, for example, the weeks before the attacks on the World Trade Center and the Pentagon. Or on that embarrassing day you

—————————————————. Given your name, a really intelligent surveillance system could even fill in the blank.

Much depends, of course, on who's in charge, us or them, and on who "them" is. A police state could exploit geographic technology to round up dissidents—imagine the Nazi SS with a GeoSurveillance Corps. By contrast, a capitalist marketer can exploit locational data by making a cleverly tailored pitch at a time and place when you're most receptive. Control is control whether it's blatant or subtle.

+   +   +

Surveillance cartography exploits diverse technologies, the most basic of which is the geographic information system, or GIS, defined as a computerized system (naturally) for storing, retrieving, analyzing, and displaying geographic data. This Spartan definition covers a variety of approaches, including overlay analysis and address matching. Around 1990 the GIS replaced the paper map as the primary medium of map analysis, and government mapping agencies like the U.S. Geological Survey shifted their focus from making maps to compiling electronic data. This change was equally apparent in higher education after GIS replaced traditional cartography as the most popular techniques course for geography majors, and disciplines like forestry and urban planning began to offer their own GIS courses.

Overlay analysis is a straightforward concept, easily visualized as a cartographic sandwich with two or more layers called coverages. As figure 1.1 shows, each coverage represents a separate topic or landscape feature, such as forest cover or soil acidity. Because the layers share a common geographic framework, the user can readily retrieve data for a particular point or define areas with a particular combination of characteristics. Overlay analysis is especially useful in exploring associations between environmental factors and in assessing the suitability of land for development.

A related GIS operation is buffering, described in figure 1.2. Planners wary of adverse effects on nearby residents or wildlife look closely at buffer zones around building sites, proposed landfills, and transportation corridors. Buffering is also helpful to emergency management officials, who need to delineate hazard zones such as those around active faults, toxic waste dumps, chlorine tanks that

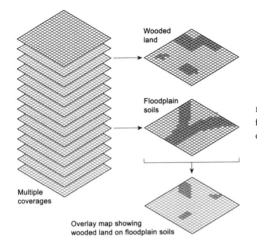

FIGURE 1.1 A GIS designed for overlay analysis relies on data organized in layers.

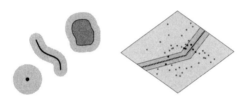

FIGURE 1.2 Buffers around point, line, and area features *(left)* are useful in evaluating the impact of hazardous or visually offensive facilities like gas pipelines or power lines *(right)*.

might rupture, and railroads and pipelines carrying hazardous materials. To explore the threat of a particularly risky shipment, an analyst defines a hazard zone around the proposed route and overlays this buffer on a detailed population map. A more advanced form of buffering called dispersion modeling can track airborne releases of radiation or lethal gasses and predict the advance of groundwater plumes fed by leaky underground storage tanks.

A more potentially invasive kind of GIS deals with street addresses, road networks, and census data. An application familiar to most Internet users is the Web site that converts an address into a detailed neighborhood map like the example in figure 1.3. The process depends upon a massive database that links addresses like "302 Waldorf Parkway" to the geographic coordinates of intersections at opposite ends of the block. Names of the city and state expedite the search by differentiating this Waldorf Parkway from all the other Waldorf Parkways in the country. The database contains each block's low and high addresses, which the GIS uses to find the specific block. If the even-numbered addresses range from 300 to 312, the lot at 302 is close to the block's low end. Assuming all the lots are equally wide and numbered sequentially, the GIS can easily calculate 302's location between the intersections and plot its position on the correct side of the street. The GIS then fleshes out the map by adding other streets in the vicinity. Knowing where you live is a starting point for probing your environment and interactions.

 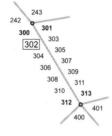

FIGURE 1.3 Online mapping services such as MapsOnUs.com and MapQuest.com plot neighborhood street maps with legible but jaggy labels *(left)* and use address ranges on opposite sides of the street to locate specific homes and businesses *(right).*

Some address-to-map Web sites also find the shortest route to another address and print out an itinerary. Because the computer knows which streets converge at each intersection, it can keep track of distances, construct and compare trial routings, and compile a list of driving directions like "Go 3.4 miles to State Highway 17, turn left onto Broad Street, and continue 0.6 miles to Main Street." A hybrid itinerary supplements these verbal instructions with small maps of the area around each turning point. Given accurate data, the system can also provide exit numbers for expressways, avoid sending motorists the wrong way down one-way streets, offer a choice between the shortest and the quickest routes, and show motels, gasoline stations, and fast-food restaurants. A particularly rich database can help motorists avoid high-crime neighborhoods and dangerous highways. Some Web sites also offer maps describing recently reported accidents and their effect on traffic—a genre of surveillance cartography I examine in chapter 6.

The prominence of commercial address-to-map Web sites supported by chain restaurants and other advertisers obscures the technology's origin as a tool for tabulating census returns. In the late 1960s, the U.S. Bureau of the Census devised a coding scheme now called TIGER (for topologically integrated geographically encoded referencing) to automatically compile block-level counts for urban areas where households received a mail-back questionnaire. The process is straightforward. An optical scanner converts each completed questionnaire to an electronic record so that a computer can match the address with the corresponding block and add the household's responses to the running tallies for various categories of age, race, and sex. Each block has a unique number as well as separate counters for each category. For example, if a home is in block 517 and its only occupants are a forty-year-old white female and an eight-year-old white male, the computer adds two each to block 517's counters for "all persons" and "white persons," and one each to the counters for "adult white females" and "white males under eighteen." Block counts are essential for congressional redistricting because the federal courts tolerate only small differences in population.

TIGER files also help retailers send catalogs and coupons to receptive homes. Ever wonder why a move to a better neighborhood triggers a different mix of junk mail? It's probably because TIGER-based address matching indicates that you're now in a more affluent census tract with a different demographic profile. Data for census tracts, which contain about four thousand persons and perhaps twenty blocks or more, include a richer variety of socioeconomic indicators than block-level data, based on the "short-form" questions the government asks all households. Age and family structure are equally relevant. If many of your new neighbors are in their fifties or sixties, ads pitching condos in Florida and long-term care insurance will be common. If most area families have young children, expect mailings that tout toys and summer camps. And direct-mail retailers who use geospatial technology to compile their own censuses from sales records can easily send you a catalog when a neighbor places an order.

Illegal aliens as well as citizens worried about privacy have little to fear from the Census Bureau. By law, the bureau cannot divulge information about individuals, even to other government agencies, and must keep their questionnaires confidential for one hundred years, after which a household's responses are of interest only to historians and amateur genealogists ferreting out ancestors. Neighborhoods are a different matter: although block-level data are comparatively innocuous tabulations pigeonholed by age, sex, race, Hispanic origin, home ownership, and residents' relationship to the household, the bureau publishes increasingly rich categorical data for block groups, census tracts, cities and towns, counties, and states. Planners and policymakers depend on these data, as do companies seeking advantageous locations for stores and restaurants. But if an area's population is so small or an individual's circumstances so unique that summary statistics might reveal sensitive information such as the income of a particular person or household, census officials suppress the summary statistics at the tract or town level. And to promote consistency, the Census Bureau has developed software that can scan tabulated data and identify numbers that, if released, would violate the nondisclosure rule. Computer-assisted enforcement of the bureau's privacy restrictions is a wise strategy that saves time and avoids charges of favoritism or malice.

Criminal records are another matter. Some states disclose the addresses of convicted sex offenders, often to the chagrin of individuals who pose little threat to the community. Although fear of rapists and child molesters is understandable, some registries include persons whose only crime is public urination. Notification practices also vary. For example, New York, which used to release information only through a pay-per-inquiry 900-number telephone service, started posting addresses and photographs of high-risk offenders on the Internet in mid-2000. Like Web sites maintained by Arizona, North Carolina, and several other states, the New York State Sex Offender Registry allows searching by name, county, or ZIP code. Intended to warn parents about pedophiles in the neighborhood, sex-offender registries depend upon self-reporting and are often error-ridden and incomplete. In addition to questioning the reliability of databases and the value of community notification, civil liberties advocates reject the posting of maps and addresses as an invasion of privacy and a barbaric form of shaming.

Less controversial are the property-tax assessment registers with which neighbors and real estate agents can find out how much you paid for your house and how many rooms it has. In states that base property taxes on fair market value, the local assessor must disclose recent sale prices and other information useful in challenging as-

FIGURE 1.4 Maryland's online Real Property System (www.dat.state.md.us/sdatweb/) invites home buyers, sellers, and residents upset about their assessments to search the database by address.

sessments ostensibly out of line with those for neighboring properties. Although publicly available assessment records are not new, a search typically required a visit to the assessor's office, which files its data by lot number, not address. Maryland and several other states now offer assessment data online by street address at a GIS-supported Web site (fig. 1.4). Several years ago, when I sold my dad's house outside Baltimore, the Maryland Department of Assessments and Taxation's online search system was especially helpful in setting an appropriate asking price. An online map showing neighboring properties even told me which property numbers to enter for other lots in the neighborhood. With a sense of selling prices in the area, I was better able to deal with real estate agents and potential buyers.

Address data also warn of explosives and toxic chemicals in your town or neighborhood. Thanks to the Emergency Planning and Community Right-to-Know Act, signed into law two years after the 1984 chemical plant disaster in Bhopal, India, industries must report the storage of dangerous materials to local officials responsible for emergency planning. The act makes this information available to individuals and environmental groups, which have started to map hazards and monitor illness. Like pedophiles, polluters lose privacy when government reveals their location.

+ + +

When dealing with agriculture, vegetation, and wetlands, geographic information systems typically treat the earth as a grid of tiny picture elements, or pixels, as in figure 1.1. Because grid cells are organized in rows similar to the parallel scan lines, or raster format, of a television screen, grid data are called raster data. By contrast, lists of points representing streets and boundaries (fig. 1.5) are called vector data because mathematicians refer to the short straight-line segments between successive points as vectors. In general, vector data are efficient for representing spot locations, streets, and census-tract boundaries, whereas raster data are tailored to analyzing soil, vegetation, and other phenomena that more or less cover the surface of an area and can be sampled from satellites. Lists of points help the computer draw routes and estimate driving distances, while grid data make it easy to compare layers or check conditions at neighboring cells.

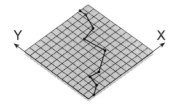

X	Y
1.0	0.0
2.5	1.0
3.0	2.7
4.4	2.5
7.7	4.7
7.2	8.0
9.4	8.6
10.5	11.0
10.5	12.0

FIGURE 1.5 Vector data represent roads, boundaries, and other linear features as lists of point coordinates. The number of points and the precision of the coordinates limit detail.

As the jig-jag streets in figure 1.3 attest, vector data sometimes cut corners quite literally by showing curved features as jaggy lines. This loss of detail is understandable for simple address maps because the substantially larger database required for more aesthetically pleasing plots would be cumbersome to process and costly to maintain. Surveyors and civil engineers find these maps useless, except perhaps for visiting a job site. Highway plans need many more points as well as more precise and reliable coordinates.

Raster data have a similar trade-off: very small pixels result in a relatively large file, and vice versa. Whether improved ground resolution is worthwhile depends on the type of analysis and its computational complexity. Satellite data with pixels 2 meters to a side can be extraordinarily valuable in urban planning and military intelligence, but the same ground resolution would give meteorologists a massive computational headache.

+ + +

Pixel size is especially important in *remote sensing*, an information technology for taking pictures of the ground from an aircraft or a satellite. Intelligence analysts want the smallest pixel possible in order to track missiles and troop movements as well as monitor more insidious threats like chemical, biological, and nuclear weapons. Spy satellites became a top priority during the Cold War, and Congress generously supported remote sensing. University researchers with government grants carried out basic research, while analysts with security clearances pored over images from the CIA's top-secret Corona satellites at the agency's clandestine National Reconnaissance Office (NRO). Promising but hardly adequate photographs from the first successful Corona mission, launched in 1959, had a ground resolution of 12 meters (40 ft.). Within eight years, a massive research and development effort had refined the pixel down to an impressive 1.5 meters (5 ft.). By contrast, civilian earth scientists had to wait until 1972 for NASA's Landsat-1 and be content with 79-meter imagery. With defense dollars driving development, intelligence satellites have always taken sharper pictures than their civilian counterparts.

The history of remote sensing during the Cold War reads like a Cinderella story in which civilian applications enjoyed the occasional hand-me-down, while only their military stepsisters were invited to the ball. As Corona demonstrated, high-resolution remote sensing was technologically feasible in 1972, but defense officials were reluctant to share classified technology that might provoke an angry Third World reaction to overhead snooping. Although Landsat-4, launched in 1982, impressed scientists with its 30-meter resolution, American civilian remote sensing lost face in 1986, when a French company, Spot Image, began selling 10-meter imagery, and American energy companies, engineering firms, and local governments became eager customers.

Defense restrictions eased markedly in the 1990s. With the Cold War over, Russia's space agency began selling 2-meter imagery on the open market and even engaged a Maryland firm to market its wares in the United States. The firm's pitch apparently impressed the U.S. Air Force, reported in 1994 as pondering a major purchase. According to the trade journal *Aviation Week and Space Technology*, U.S. intelligence imagery was too detailed for Air Force needs, and the Russians offered a better combination of resolution and price than any other source. America recovered its prominence in civilian satellite surveillance in 1999, when Space Imaging, a Colorado company, launched Ikonos-2 and offered to sell or lease 1-meter imagery to all comers, domestic or foreign. Almost all, that is—Washington strongly discourages the sale of high-resolution satellite imagery of Israel, and during the 2001 Middle Eastern campaign, the government thwarted enemy and media hopes by buying exclusive rights to Ikonos imagery of Afghanistan.

If you're a small state surrounded by hostile neighbors, 1-meter satellite imagery doesn't afford much privacy. Iraq and Syria would pay millions to learn where to invade or bomb, or how best to repel an Israeli attack. But as the Ikonos snapshot of the Washington monument in figure 1.6 illustrates, an eye in space records more than targets and fortifications. Roads and buildings are especially obvious, but it's also easy to identify trees, vehicles, and pathways. And the local assessor can readily recognize a new swimming pool in your backyard. Intelligence satellites have even sharper eyes: various estimates suggest that pictures from Corona's most advanced successors have a resolution of roughly 3 inches.

High-resolution snapshots from space are not the only kind of overhead surveillance. Aircraft flying at a few thousand feet capture comparable detail with less sophisticated cameras to the delight of thrifty mapmakers, planners, and civil engineers, who rely largely on conventional aerial photos much like those shot in the 1930s. And airplanes as well as satellites can carry a wide variety of sensors, ranging from optical cameras with photographic film to the sophisticated synthetic aperture radar system with which the space shuttle *Endeavor* mapped most of the world's terrain in little more than a week. As chapters 2 and 3 discuss, electronic imaging systems address a broad and growing array of military and commercial needs.

FIGURE I.6 Portion of first Ikonos image of Washington, D.C. The Washington Monument is at the lower left, and the longer, darker shape is the monument's shadow. Deciduous trees were still in full foliage when Ikonos-2 took this 1-meter image on September 30, 1999. For additional examples and further details, see the Space Imaging Web site (www.spaceimaging.com). Courtesy of Space Imaging.

A pivotal advance was the infrared scanner, sensitive to light our eyes can't see yet able to distinguish camouflage from natural vegetation as well as pinpoint distressed crops. In the 1970s, thermal scanners extended human vision still further by measuring heat loss and soil moisture. Advances in image processing software help researchers and intelligence analysts classify vegetation and search for suspicious objects or operations, such as missile launchers or indoor marijuana plantations. Change detection, an important aspect of cartographic surveillance, uses the overlay techniques of GIS to compare imagery for different dates.

+   +   +

Remote sensing and GIS rely heavily on a third, much newer electronic technology, the global positioning system, or GPS. Developed by the Defense Department to help troops and missiles find themselves on maps, GPS is an efficient technique for entering new information—field observations as well as remotely sensed data—into a GIS. A classic example of post–Cold War technological trickledown, the military's constellation of GPS satellites now supports a host of civilian applications, including boundary surveys and highway navigation. In archeology, geology, and soil science, for instance, reliable, low-cost GPS receivers have become a standard tool for collecting and mapping field data. And in the consumer electronics market, GPS is a promising enhancement to the personal computer, the personal digital assistant (like the Palm handheld), and the car radio. Once an expensive option for car buyers, the dashboard navigation system will no doubt replicate the success of the FM radio and the CD player and make road maps as we know them as obsolete as 8-track tapes and 45 rpm records.

Equally adept at tracking vehicles, employees, adolescents, and convicted criminals, GPS is very much a surveillance technology, with credible threats to personal privacy. Just ask the former clients of Acme Rent-a-Car, a Connecticut firm that tracked its vehicles by satellite and fined customers for exceeding 79 MPH.

GPS calculates location by comparing time signals from several satellites, each with a direct line of sight to the receiver. Each satellite broadcasts a signal traveling at the speed of light but requiring a measurable time to reach the ground. At a velocity of 186,000 miles per second, the signal takes 0.06774 seconds to reach a GPS receiver directly beneath a satellite at an altitude of 12,600 miles, and a bit longer to cover measurably greater distances to points elsewhere within the satellite's footprint. Although the satellites are moving, the receiver can estimate the satellite's location at the time of transmission. Because the signal encodes the time at which it was broadcast, the receiver can determine the elapsed time from satellite to ground, convert this time to a distance, and compute a circle representing all locations on the surface that far from the satellite. Signals from other satellites yield additional circles, which intersect at the receiver's location (fig. 1.7). And because the triangulation is three-dimensional, the GPS estimates elevation as well as latitude and longitude.

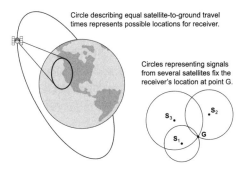

Circle describing equal satellite-to-ground travel times represents possible locations for receiver.

Circles representing signals from several satellites fix the receiver's location at point G.

S₃ S₂ S₁ G

FIGURE 1.7 GPS uses three-dimensional triangulation based on intersecting circles describing travel time from space to ground for signals from several satellites. Each circle describes a range of locations equidistant from one of the satellites, and the circles' point of intersection represents the receiver's location.

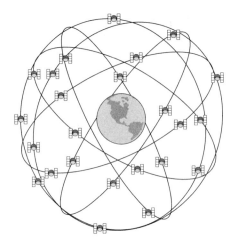

FIGURE 1.8 The GPS constellation consists of twenty-four satellites in six orbital planes spaced evenly, 60 degrees apart, and inclined 55 degrees above the plane of the equator. In turn, the four satellites in each plane are evenly spaced in a circular orbit with an altitude of 20,200 kilometers (12,600 stat. mi., or 10,900 naut. mi.). The pattern is easily predictable, with each satellite circling the earth twice a day and tracing the same ground track across the earth below every 11 hours, 58 minutes. Adapted from *NAVSTAR GPS User Equipment Introduction*, public release version (Alexandria, Va.: U.S. Coast Guard Navigation Center, September 1996): p. 1-3, fig. 1-2.

Although GPS receivers can compute location on the fly, accuracy improves when a stationary receiver takes multiple readings, from different positions along each orbit. Because it's also helpful to integrate signals from more than three satellites—more signals mean less uncertainty—the Air Force, which runs the system, maintains a constellation of twenty-four satellites (fig. 1.8), at least four of which should be visible at any time from any place on earth.

Because taxpayers paid over $10 billion to build the GPS system, letting them reap some of its benefits is good politics. And granting equal access to users worldwide creates goodwill as well as an overseas market for American electronics. But goodwill has too high a price, the Defense Department argued, when enemies and allies have equal access. So the White House compromised with a policy of deliberate blurring called Selective Availability (SA). Under a two-tiered system of signals, only military users received all the information needed to estimate location with great precision. And because an enemy might retaliate electronically, the Precise Positioning Service (PPS) gave soldiers and GPS-guided cruise missiles a signal that was less readily jammed than the Standard Positioning Service (SPS), designed for nonmilitary users. Thus, a military receiver could nearly instantly estimate horizontal location to within about 10 meters, while its civilian counterpart might be off by 100 meters or more.

Selective Availability proved a costly and needless inconvenience. During the 1991 Gulf War and the 1994 Haiti campaign, for example, a shortage of the more expensive PPS receivers forced the military to cut down the error injected into the SPS signal. What's more, civilian users willing to wait a bit or link to a precisely located ground station could readily reduce uncertainty to 1 meter, which is sufficiently accurate for drivers and pedestrians with electronic street maps (fig. 1.9). On May 2, 2000, the Defense Department surrendered to pressure from the domestic electronics industry and stopped blurring civilian GPS signals. Dennis Milbert, a top official at the National Geodetic Survey, compared the improvement to a football stadium in which, "with SA activated, you really only know if you are on the field or in the stands [whereas] with SA switched off, you know which yard marker you are standing on." Even so, the Pentagon remains in control. As President Clinton noted in a statement released on May 1, the military can "selectively deny GPS signals on a regional basis when our national security is threatened."

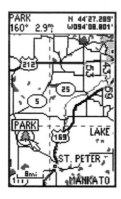

FIGURE 1.9 The ability to zoom in and out allows handheld GPS receivers like the Garmin 12MAP *(left)* to display small but useful maps *(right)*. © Garmin Corporation. Reproduced with permission.

Privacy concerns arise principally with proposals to integrate GPS with other technologies. Connect a portable ground station to a cell phone, for instance, and it becomes an instrument for tracking employees, children, and parolees. Although GPS surveillance can be quite benign—an efficient way to track and dispatch taxis, for example—proposals to clamp a tracking device around a subject's ankle or implant it beneath the skin in a microchip have sinister overtones akin to stalking and branding. Add the possibility of administering pain if a linked GIS observes a child or ex-convict entering forbidden territory, and the scenario is instantly Orwellian. It's easy to see how punitive electronic tracking might work. The GPS transmits the subject's coordinates to a computer with a detailed geographic database describing no-go areas, diligently delineated with vector data. If the computer detects a first-grader playing near a garbage dump or busy street, the system intervenes with a warning, by pinch or pager. I can see how some child safety experts might approve an auditory warning device IF it's reliable. (My big "if" is deliberate; the computer could be down or its database faulty.) I also see why civil libertarians would object vigorously to remote spanking that could, quite literally, follow a child into adolescence and beyond. But I much prefer my own reasons for never using a child tracker: if parts of the community must be off limits, the parent should teach the child to understand and recognize hazards. If my daughter were too young to appreciate danger, she wouldn't be away from home without adult supervision.

In some circumstances, personal tracking might well be the lesser of several evils. Consider, for example, the alternatives to permanent incarceration of convicted sex offenders. Compared to minimally supervised parole, which could pose a threat to the community, and public outing, likely to inflict emotional stress and undermine rehabilitation, GPS-based monitoring—IF it works— seems a good choice. And consider the senior citizen for whom a tracking device might be—IF it works and monitors relevant vital signs—a suitable substitute to confinement in a nursing home or an adult care center. Even so, both cases admit other solutions as well as circumstances that readily rule out cartographic surveillance. Whatever the ethics and pragmatics of location tracking, the unintended consequences of GPS call for skeptical awareness of our brave new globe and its plausible threats to personal privacy.

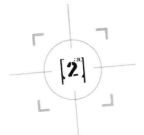

Overhead Assets

Overhead surveillance has three histories: two focusing on
the distinctly different technical challenges of aviation and
photography and a third encompassing the politics and in-
stitutions of a hybrid defense-oriented technology with commercial ambitions. Advances in fluid mechanics, rocket
propellants, and inertial navigation allowed aircraft and artificial satellites to overcome gravity and atmospheric drag,
while cameras and scanners, as extensions of the human
eye and the artist's canvas, have historical roots in optics,
photochemistry, and radar. Although their respective narra-
tives are still evolving, space flight and electronic imaging
have attained impressive plateaus. By contrast, the hybrid
technology's history is less complete but more intriguing.
And as this chapter illustrates, the intelligence commu-

nity's "overhead assets" play a critical yet largely unsung role in national defense and world peace.

+   +   +

Like most jargon, "overhead assets" is mildly ambiguous: although aerial reconnaissance is almost always an asset, the observer's eye or camera's lens need not be directly overhead. The French army, which used aerial scouts as early as 1794, tethered their balloons to protect observers and assure timely reports. A mile or more from enemy positions, aeronauts with telescopes could assess an adversary's strength, help artillery officers improve their aim, and direct ground attacks toward weak points along the opponent's line. Because photography would not be invented for several decades, aerial observers made sketches and annotated existing maps, which were conveniently dropped to intelligence officers below. Although the daguerreotype became practical in the late 1850s, long exposure times and cumbersome glass plates limited its military use to anticipatory mapmaking on calm, clear days. The Union army, which deployed a few balloons during the Civil War, discovered another impediment as the war widened: the difficulty of moving aeronauts and their apparatus quickly to new locations.

Airplanes, introduced to battle during World War I, solved the transportation problem and allowed true overhead reconnaissance. As long as a pilot stayed above 18,000 feet, he was generally immune to enemy gunfire. Attack planes designed to overtake bombers and outmaneuver the opponent's fighter aircraft were another matter. Although civilian aerial photography as it evolved in the 1920s and 1930s relied on slow, evenly paced, back-and-forth flying along adjoining, closely spaced flight lines, reconnaissance pilots had to get in and out quickly. Because leisurely flight over enemy terrain was rarely an option, their most informative shots seldom looked directly downward.

Specialized cameras let reconnaissance planes and bombers make the most of their time over hostile territory. During World War II, American engineers devised a rigid frame with three cameras, one pointing directly downward and the others aimed away from the vertical, to the left and right (fig. 2.1). The result was a trio of slightly overlapping pictures: a more detailed vertical shot centered near the nadir point, on the ground directly below the camera, and a pair of oblique views looking outward toward the horizon. Although scale varied widely on these oblique shots, trained photointerpreters could examine a vast area and transfer potential targets to existing maps. Multi-lens cameras were especially efficient for rapid systematic small-scale mapping because the wider nadir swath required fewer flight lines. To assure complete coverage, an electrical or mechanical device snapped all three shutters simultaneously, at a constant interval, while other motors advanced the film automatically. During the mid-1940s, the Army Air Force used three-lens photography to compile a set of small-scale flying charts covering 16 million square miles (41 million km²).

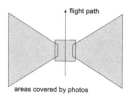

cameras

flight path

optic axes of cameras

areas covered by photos

FIGURE 2.1 The three-lens aerial camera *(left)* captured a vertical image surrounded by a pair of oblique views *(right)*.

Oblique viewing helps satellite sensors respond to emergencies with timely images. Commercial satellites like SPOT and Ikonos offer nearly worldwide coverage by orbiting the earth several times a day in a geometric plane that moves steadily westward so that the satellite always crosses the equator at the same time, usually in late morning. As the satellite circles overhead, its scanner examines a narrow ground swath 10 to 200 kilometers wide and doesn't pass directly overhead again for another twenty days or so. Swath width and revisit time reflect the resolution of the satellite's scanner as well as its altitude or orbit. The early Landsats provided 79-meter resolution from an altitude of about 920 kilometers along a swath 185 kilometers wide revisited every eighteen days. By contrast, Ikonos offers 1-meter resolution along a swath only 11 kilometers wide from an altitude of 681 kilometers, but cannot retrace the same nadir track for several months. But by tilting the sensor outward (fig. 2.2, left), Ikonos can advertise revisit times of 2.9 and 1.5 days, respectively, for resolutions of 1.0 and 1.5 meters. Although the larger tilt angle required for a shorter revisit time undermines

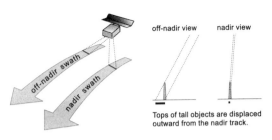

off-nadir view   nadir view

Tops of tall objects are displaced
outward from the nadir track.

FIGURE 2.2 Off-nadir views to the left and right of the nadir swath *(left)* reduce revisit time but displace features outward from the nadir point *(right)*.

resolution, off-nadir viewing is an essential element of high-resolution remote sensing.

Off-nadir images contain distortions common to the periphery of perfectly vertical aerial photographs. When the scanner or camera is not directly overhead, the image shows only the near side of tall structures like the Washington Monument (toward the lower left in fig. 1.6). Similarly, the top of a perfectly vertical obelisk will appear farther from the photo's nadir point than the monument's base (fig. 2.3, left). What's more, the distance between the tops of two obelisks, equal in height but at opposite corners of the photo, will be measurably greater than the distance between their bases. This phenomenon is called relief displacement because higher features, especially near the edges of the photo, are displaced farther outward from the center of the photo than lower features. Because of relief displacement, air photos distort mountains and tall buildings and should not be used to measure horizontal distances, especially in rugged terrain.

Relief displacement proved more an asset than a liability. A pair of overlapping aerial photographs taken from different camera positions (fig. 2.3, right) can be viewed in stereo, with each eye seeing the landscape from a slightly different angle—the same principle employed in the nineteenth-century stereopticon and late-twentieth-century virtual reality. Able to examine land cover in three dimensions, photointerpreters quickly learned to distinguish two-story houses from low-profile chicken coops and identify other features not readily apparent on a single photo. What's more, photogrammetrists devised instruments for measuring relief displacement,

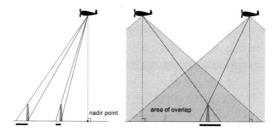

FIGURE 2.3 Because an air photo is a perspective view, the top of a vertical monument is displaced outward from its base *(left)*. On overlapping air photos the joint displacement *(right)* of an object's two images promotes stereovision and provides measurements useful in calculating differences in elevation.

calculating elevations, drawing contour lines, and replotting features in their horizontally correct positions. During the 1930s photogrammetry became the preferred technique for making topographic maps, and three decades later photogrammetrists devised an efficient method for removing relief displacement from aerial photos.

Satellites with dual off-nadir scanners, one aimed forward along the ground swath and the other looking aft (fig. 2.4, left), can mimic the overlapping pairs of air photos used in stereo compilation. A scene captured twice, from different viewing angles, allows stereo viewing and provides joint displacements (fig. 2.4, right) for estimating elevation differences. Imaging software that matches features on the fore and aft views can automatically compile an elevation map, which can be plotted as a three-dimensional terrain diagram, con-

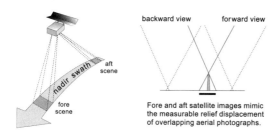

FIGURE 2.4 Fore and aft views *(left)* along a satellite's nadir swath afford joint displacements *(right)* that allow stereo viewing and automatic compilation of elevation maps.

toured like a typical topographic map, or used to filter out all relief displacement. Displacement-free images, called orthophotographs, are also possible when overlapping fore and aft scenes are captured on a neighboring, off-nadir swath.

+   +   +

Like most technologies, remote sensing boils down to a few fundamental trade-offs. Increase altitude, and resolution suffers. Increase resolution with a larger, heavier camera, and you need more fuel, a bigger engine, or a radically different airplane or satellite. In wartime, altitude becomes a conflicted matter of safety and need to know—accurate intelligence is crucial but flying too low can get you shot down. During the Cold War, intercontinental ballistic missiles (ICBMs) and nuclear warheads highlighted another trade-off by vastly increasing the cost of knowing what the enemy was up to as well as the consequences of not knowing and doing nothing. Aware that the Soviet Union now had atomic weapons as well as long-range missiles, the United States needed to monitor its key rival's nuclear activities and missile sites.

So great was the cost of inadequate intelligence that the government launched a crash program of top-secret spy satellites to complement and eventually replace sporadic overflights by U-2 reconnaissance planes, designed for long flights at 70,000 feet, well out of reach of Soviet fighters. First used over Russia in 1956, the one-person spy plane had replaced a less subtle program of unmanned and largely unreliable reconnaissance balloons, which occasionally provoked diplomatic protest. Overflights of the USSR halted abruptly on May 1, 1960, when a surface-to-air missile shot down a U-2 near Sverdlovsk, and the Russians put the pilot on trial as a spy. Fortunately for the CIA, the U-2 was not its only option. On August 18 the first successful Corona satellite, launched as Discoverer-14, photographed more Soviet territory during its single day in orbit than the twenty-four U-2 missions of the previous four years.

Equally impressive was the Air Force's mid-air snagging of the Discoverer-14 recovery capsule carrying 3,000 feet of 70 mm film. Unlike Landsat and its technological cousins, which beam their pictures down to earth electronically, the early Corona satellites were giant disposable cameras. But instead of the entire unit having to be returned to Kodak for processing, the satellite ejected its film in a gold-plated recovery capsule that resembled an oversize kettle-drum. De-orbited by a retro-rocket and dropped over the Pacific Ocean with a parachute, a Discoverer capsule could float for two days—sufficient time for recovery by a Naval team that knew what to look for and where to look. And if the Navy couldn't retrieve it, no one would—a salt plug in the flotation unit would dissolve, and the capsule would fill with water and sink. However clever, water recovery was only a backup: ideally a huge C-119 or C-130 transport plane equipped with a trapeze-like sling would stalk the slowly descending recovery capsule, seize its parachute, and reel the catch into the cargo bay.

Discoverer was a clever double entendre: as a cover name it implied space exploration and biomedical research, but the CIA and Air Force officials running the program focused on discovering airfields, launch pads, and uranium plants. Subterfuge was essential because the powerful two-stage rockets needed to loft a Volkswagen-size satellite into space would surely be noticed by the news media as well as curious residents of Santa Barbara, California, near the launch site at Vandenberg Air Force Base. Although Corona was "deep black"—intelligence community lingo for absolutely top secret—many of the program's goals and operations were openly reported in the trade magazine *Aviation Week*.

A dozen disappointments preceded Discoverer-14's success—rockets failed, satellites spun out of control, and cameras malfunctioned. Launched just one week before Discoverer-14, Discoverer-13 proved a turning point: recovery of the nonphotographic diagnostic mission's reentry capsule, which splashed down 330 nautical miles from Hawaii, marked the first successful retrieval of an object sent into space. But three more failures intervened before the second successful air recovery of film from Discoverer-18, launched on December 7, 1960. It took another year for engineers to work out bugs. Although only six of the sixteen photographic satellites launched in 1961 returned film, seventeen out of twenty Corona missions in 1962 were successful.

Efficient coverage demanded careful coordination of camera and orbit. A slowly rotating near-polar orbit assured multiple ground swaths across the Asian heartland, where the Soviets situated their most secret missile sites and atomic laboratories. The orbit was not a circle but a slightly eccentric ellipse, which varied in altitude from more than 800 kilometers to less than 180 kilometers, to bring the satellite in sufficiently low so that Corona's 5-foot-long camera with a 24-inch-focal-length lens could take pictures that mimicked a resolution of 12 meters. Instead of snapping a shutter, advancing the film, and taking another snapshot a short while later like the cameras used in conventional aerial photography, Corona captured a panoramic view of the scene below on a continuously moving strip of film. As the satellite moved along overhead, the film recorded a ground spot moving perpendicular to the nadir track and reaching outward 35 degrees from the vertical on both sides to capture a swath at least 250 kilometers wide. Continuous exposure avoided gaps and wasteful overlap, and the carefully programmed camera conserved film by turning itself on and off at appropriate points. Although continuous strip panoramic photography was not new, Discoverer-14 set a record by imaging 1,650,000 square miles (4.3 million km²) of Soviet terrain in a single day.

The first Corona camera system was the KH-1, where KH stands for Keyhole, the government's code name for top-secret satellite reconnaissance. Between 1959 and 1962, three new models reflected numerous improvements. KH-2 introduced a more accurate film-advance and scanning mechanism, and refined the resolution to 8 meters. KH-3 introduced a faster lens, which allowed slower, finer-grain film and sharper enlargements. KH-4 extended mission life to six or seven days with additional film and pioneered satellite stereo imaging with two cameras, one aimed 15 degrees forward along the nadir swath and the other pointed 15 degrees aft. Bigger in this case also meant better: according to recently released records, the CIA rated twenty-one of the 9-foot-long camera's twenty-six missions as successful.

Subsequent advances were equally impressive. The KH-4A imaging system, first used in August 1963, added a second recovery capsule, or "bucket," and increased the total film load to 32,000 feet, which allowed missions as long as fifteen days. Other improvements included a resolution of 2 to 3 meters (7 to 10 ft.), depending on altitude, stereo coverage of 18 million square miles (47 million km²) on a typical mission, and a camera malfunction rate of only

4 in 52. The 12-foot-long KH-4B, introduced in September 1967, lengthened mission life to nineteen days, offered greater flexibility in type of film and exposure, and honed routine image resolution down to 6 feet (less than 2 m). In addition to the forward and aft panoramic cameras for stereo imaging, the KH-4B included three types of secondary cameras, which recorded scenes described schematically in figure 2.5. An index camera with a square format captured small-scale terrain images useful for relating and indexing paired sets of panoramic images. Each panoramic camera had its own pair of horizon cameras—one aimed to the left of the nadir swath and the other pointing to the right—to establish altitude and scale. Two stellar cameras, pointing above the horizon to the left and right, recorded the positions of visible stars. Celestial signposts were important to Defense Department mapmakers, who used astronomic observations to calculate latitude and longitude. Clandestine pictures might help the CIA keep tabs on enemy armaments, but if the United States ever needed to retaliate, its guided missiles had to know each target's exact location.

Corona imagery remained sequestered in National Reconnaissance Office vaults until 1995, when President Clinton authorized the release of "scientifically or environmentally useful . . . historical intelligence imagery." Although the official intent was to provide baseline data for assessing environmental change, declassification of more than 800,000 frames of top-secret photography made further denial of American capacity for overhead surveillance pointless, especially when Russian space entrepreneurs were hawking equally sharp imagery to cost-conscious Air Force officials.

Declassification was an opportunity to brag. Engineers who had worked on the rockets, satellites, and cameras could speak openly about their achievements, while intelligence experts praised Corona's accomplishments. By monitoring weapons tests and identifying all Soviet missile sites, defensive and offensive, NRO interpreters had exposed the "missile gap" of the late 1950s as a groundless fear. Satellite intelligence proved especially useful in October 1962 during the Cuban missile crisis, a thirteen-day standoff between Soviet Premier Nikita Khrushchev, who had threatened to bury us, and President John F. Kennedy, who knew the Russians were poorly prepared for a major offensive. In tracking the deployment of

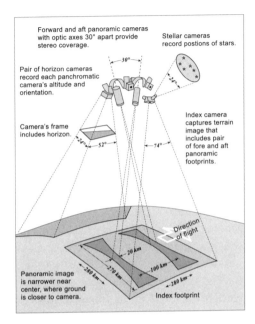

Forward and aft panoramic cameras with optic axes 30° apart provide stereo coverage.

Stellar cameras record postions of stars.

Pair of horizon cameras record each panchromatic camera's altitude and orientation.

Camera's frame includes horizon.

Index camera captures terrain image that includes pair of fore and aft panoramic footprints.

30°

52°   74°

20 km

280 km   270 km   100 km

280 km

Direction of flight

Panoramic image is narrower near center, where ground is closer to camera.

Index footprint

FIGURE 2.5 The KH-4B imaging system supplemented its fore and aft panoramic cameras, which provided stereo coverage, with an index camera, a pair of stellar cameras, and a pair of horizon cameras for each panoramic camera. Footprints describe areas on the ground covered by images framed for storage and use. This schematic diagram shows only one of the KH-4B's four horizon cameras. Redrawn from figure 8 in National Reconnaissance Office, *The KH-4B Camera System*.

atomic weapons and submarines as well as monitoring Russian assistance to China, Cuba, and its Middle Eastern allies, satellite imagery also demonstrated the feasibility of checking up on rivals and verifying arms-control treaties. Corona not only became the Defense Mapping Agency's key resource for compiling medium-scale maps but also provided basic observations for a long-overdue revision of the worldwide geodetic system, which tells cartographers where to draw meridians and parallels. Less well known is the U.S. Geological Survey's use of classified satellite imagery to revise its 1:250,000, medium-scale topographic maps as well as identify out-of-date 1:24,000 quadrangle maps. Although revision of large-scale maps required conventional aerial photography, straightforward comparison of existing maps with Corona imagery helped USGS focus its efforts on comparatively needy quadrangles.

+   +   +

Corona's final launch in May 1972 provided a convenient cutoff for further official disclosures. But even though the NRO still sequestered post-Corona imagery, journalists specializing in space technology and strategic intelligence offered revealing glimpses of the more sophisticated systems that followed. Foremost among them are John Pike, director of GlobalSecurity.org and formerly the satellite intelligence guru at the Federation of American Scientists, and Jeffrey Richelson, the author of several books on satellite intelligence and the history of electronic spying. Pike, who worked as a science writer and political consultant before joining the FAS, is a respected critic of government space programs and an ardent supporter of arms limitation. Richelson, who holds a Ph.D. in political science, taught government at American University before becoming a full-time author and consultant. Sought out by news reporters whenever breaking stories involve spy satellites, Pike and Richelson were key sources for William E. Burrows, whose 1987 exposé *Deep Black: Space Espionage and National Security* is a surprisingly accurate unofficial history of Corona and Keyhole.

All three authors have relied heavily on technical documents from official sources as well as public testimony before congressional committees looking into arms limitation. The White House had backed test-ban and arms-reduction treaties for over a decade—Richard Nixon helped initiate the Strategic Arms Limitation Treaty (SALT), and President Jimmy Carter called nuclear arms reduction his "most cherished hope"—but skeptics in Congress demanded an effective means of verification, including overhead surveillance. Although the intelligence establishment promised diligent photographic reconnaissance, neither the CIA nor the Defense Department would confirm publicly their use of spy satellites.

The curtain of secrecy remained intact until October 1978, when Carter, speaking to NASA employees at the Kennedy Space Center, pledged that "photoreconnaissance satellites . . . make an immense contribution to the security of all nations [and] we shall continue to develop them." Unofficially of course, Washington knew we had spooks in space. Well-placed but appropriately guarded informants, probably with approval from their superiors, confided in lawmakers and journalists without betraying sensitive details. Added reassurance came from the apparent success of telecommunications and meteorological satellites, Landsat, and the space shuttle, first flown in space in April 1981.

Espionage trials brought further revelations. In 1977, for instance, CIA employee Wilham Kampiles sold a top-secret technical manual to a Russian agent for $3,000. The manual described the operation and limitations of the KH-11 imaging system, employed in a Corona successor. In addition to further confirming Russian knowledge of American spy satellites, Kampiles's November 1978 trial highlighted the conviction a year and a half earlier of Christopher Boyce and Andrew Lee, who had sold secrets about U.S. spy satellites developed by Boyce's employer, the TRW Space Systems Group. In 1980 Boyce further embarrassed the government by escaping from a federal prison and remaining at large for ten months.

According to Pike and Richelson, the Russians clearly knew what the CIA was doing. The two authors testified as defense witnesses at the 1985 trial of Samuel Morison, a former Navy intelligence officer charged with espionage. Morison had sent top-secret satellite photos to *Jane's Defence Weekly*, a prominent British military magazine, which published three of them in an August 1984 issue. The high-resolution images afforded a detailed view of a nuclear-powered aircraft carrier under construction at a Soviet shipyard on the Black Sea. Although motivated by journalistic zeal—he was one of *Jane's* American editors—Morison had broken the rules. Even so, his disclosure was hardly as damaging as prosecutors claimed. The pictures "just didn't really tell me anything that I didn't know," Richelson told the *Washington Post*. "I don't think they provide any new information. Therefore, I don't think it's of any value." Despite evidence that Morison's leak was neither significant nor criminal, a federal court sentenced him to two years in prison.

Unofficial disclosures abound in Richelson's paper "The Keyhole Satellite Program," published in the June 1984 issue of the *Journal of Strategic Studies*. In 1960, the White House set up the super-secret National Reconnaissance Office to manage satellite spying as well as mediate disputes between the CIA and the Air Force. In 1963, the NRO initiated a second generation of space reconnaissance by launching the first satellites with KH-5 and KH-6 systems. The KH-5, a "surveillance" system designed to cover a relatively wide ground swath, carried more film than the KH-4 and extended the typical mission to twenty-three days. By contrast, the KH-6, a "close-look" system intended for sharper pictures of smaller scenes, strengthened resolution to 6 feet (2 m) with an altitude as low as 76 miles (122 km). In 1966, the first KH-7 and KH-8 missions heralded a third generation of surveillance and close-look satellites. According to Richelson, both series operated at generally lower altitudes than their predecessors, and when orbited as low as 82 miles (132 km), the KH-8 could refine its resolution to a remarkable 6 inches (15 cm). A fourth generation emerged in 1971, when the NRO flew the first KH-9 craft, a 30,000-pound, 50-foot-long satellite unofficially known as Big Bird. With an enormous payload, four recovery modules, and missions lasting six months or more, the KH-9 combined conventional film photography for taking sharp, 6-inch-resolution close-ups with a TV-like imaging system able to transmit pictures to earth electronically.

Big Bird's days were numbered. In 1971, the NRO also launched the first KH-11, which replaced film drops with a digital multispectral imaging system similar to but sharper than the scanner orbited a year later on Landsat-1. Not burdened by film or recovery capsules, the KH-11 craft embarked on missions of years, not months. One of Richelson's sources was aerospace journalist Philip Klass, whose 1971 book *Secret Sentries in Space* revealed plans to develop near-real-time satellite reconnaissance. Another was a 1972 news note in *Aviation Week and Space Technology* announcing the imminent selection of TRW to "develop a new generation of reconnaissance satellites [that will] permit real-time photo reconnaissance by means of synchronous data relay satellites." As Richelson explained, when a KH-11 satellite was not within range of a ground station, it would relay imagery to Washington through a telecommunications satellite. What's more, off-nadir cameras allowed coverage anywhere in the world at least once a day, clouds and darkness not withstanding. With a pair of KH-11 satellites in carefully spaced sun-synchronous orbits, NRO intelligence analysts could inspect sensitive areas several times a day. And because the agency usually replaced its satellites before they failed, lack of coverage was rare. From 1977 through 1983, for instance, at least one KH-11 was operating overhead on all but seventeen days.

The KH-11's filmless photography depended on tiny, light-sensitive semiconductors called charged-coupled devices (CCDs). Invented at Bell Laboratories in 1970, the CCDs were arranged in rows and columns to capture the scene below as an array of pixels, each recording the intensity of received light as an electrical charge. The imaging system polled its pixels systematically, row by row, and converted their charges to numbers for virtually error-free transmission. Received in Washington nearly instantaneously, the satellite's digital pictures could be manipulated on a computer to improve contrast and stored electronically for ready comparison with later images.

Impressed with the KH-11's rapid read-out and extraordinary longevity, Richelson says little about its resolution, deemed "greater than the KH-9 [but] inferior to the KH-8." More intriguing was its successor, the KH-12, believed to include thermal imaging for nighttime spying as well as radar imaging to penetrate clouds. First launched in 1986, the KH-12 weighed 14 tons (18 tons in later versions). In addition to a massive camera, similar to the Hubble Space Telescope in its use of large mirrors, the KH-12's satellite carried up to 7 tons of fuel, which powered the rockets used to adjust its orbit. John Pike, who describes the KH-12 as a "space telescope with a rocket," notes the importance of moving closer to interesting areas and evading antisatellite weapons. Pike also highlights the difficulty of keeping salient details secret by calling the much publicized Hubble Space Telescope "an unclassified version of the KH-12."

Although the NRO remains mum about the KH-12's existence, much less its resolution, Pike and other experts accept a conveniently rounded estimate of 10 centimeters (3.9 in.). A calculation by electronics journalist John Adam suggests this number might be conservative. By relating the space telescope's camera to the KH-11's altitude, Adam computed a ground resolution of 7.16 centimeters —a mere 2.8 inches. Although 4-inch resolution seems more likely, not even 3-inch resolution would let an NRO analyst read news paper headlines or license plate numbers, as enthusiastic reporters occasionally claim. In fact, Pike used images of both objects at different resolutions to show that not even 1-centimeter resolution would make normal license plates and headlines readable from space.

Pike has demonstrated that 10-centimeter resolution is quite sufficient for most intelligence needs. Reproduced in figure 2.6, his array of images shows a parking lot viewed from above with ground resolutions of 10, 25, 50, and 100 centimeters. At 10 centimeters, an

FIGURE 2.6 Images of a parking lot viewed with resolutions of 10, 25, 50, and 100 centimeters. Courtesy of John Pike and the Federation of American Scientists.

analyst can describe individual vehicles and even judge drivers' skill in parking between the lines—the large truck at the lower left is too long for its space and partly over the line. At 25 centimeters, the lines are hazy and cars are barely distinct from vans and pickup trucks. At 50 centimeters, identification is difficult at best, although some lines are marginally visible. And at 100 centimeters, the same resolution as the Ikonos image in figure 1.6, size and pattern suggest vehicles in a parking lot, but an analyst would need a wider scene to be certain. It's clear, though, that 10-centimeter intelligence imagery is far more detailed than the 1-meter imagery available from Space Imaging and its competitors.

+   +   +

NRO analysts and their collaborators in the aerospace industry are haunted by a basic question in image intelligence: How fine a ground resolution is sufficient? Because the answer depends on what the analyst is trying to discover, a committee of scientists and intelligence experts developed the National Image Interpretability Rating Scales. Focused on usefulness, NIIRS assigns images and imaging systems an integer rating between 1 and 9, which corresponds to a range of "ground resolved distances," starting with "over 9.0 m" for NIIRS 1 and continuing through "less than 0.10 m" for

NIIRS 9. Each rating includes a list of identifications, descriptions, or differentiations that become possible at that level of resolution. According to the ratings, Ikonos and KH-12 imagery are radically different. One-meter Ikonos imagery, with the capabilities of NIIRS 5 (ground resolution between 0.75 and 1.2 m), lets users identify specific types of surface-to-surface missiles, distinguish between vehicle-mounted and trailer-mounted radar, differentiate steam and diesel locomotives, and classify rail cars by type (flat cars, tank cars, box cars, gondolas, and so forth). And as with slightly less refined pictures, viewers can count vehicles and detect rail yards, airstrips, helipads, radar installations, and missile silos with their doors open. What Ikonos users can't do is identify the spare tire on a medium-size truck (NIIRS 6, with 0.40–0.75 m resolution) or count individual rail ties (NIIRS 7, with 0.20–0.40 m resolution). By contrast, a trained intelligence analyst with KH-12 imagery, rated NIIRS 8 (0.10–0.20 m), can see a truck's windshield wipers or an airplane's rivet lines. If the analyst knows what to look for, the KH-12's superior ground resolution might mean the difference between merely seeing a truck and identifying its contents.

Because image intelligence focuses on detecting change, 1-meter satellite imagery is often more informative than its NIIRS 5 rating suggests. A new railway spur or clearing, for instance, could signify a new missile site or weapons factory. And a suspicious accumulation of large vehicles might presage an imminent attack. As John Pike observes, "if a picture is worth 1,000 words, two pictures are worth 10,000 words." Look for even bigger word counts once Space Imaging begins marketing the 0.5-meter imagery approved by the National Security Council in late 2000.

However sharp and revealing, visible imagery is nearly useless on a cloudy day. To compensate for clouds as well as darkness, the NRO turned to synthetic aperture radar (SAR), which mimics a flash camera by generating its own energy. Instead of light, an SAR antenna transmits pulses of microwave radio waves, which bounce back to the spacecraft (fig. 2.7). SAR differs from ordinary radar because its antenna is in motion. The distance covered between a pulse's transmission and the return of "backscattered" radiation imitates the much longer antenna needed to recognize revealing variations in terrain and surface objects. Although estimates of

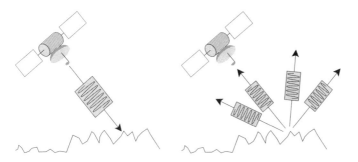

FIGURE 2.7 Radar imaging systems transmit pulses of electromagnetic energy, which are "backscattered" in various directions. Computers on the ground use the relative strength of radar echoes to reconstruct the terrain as well as the position and general shape of surface objects.

ground resolution are vague, space-based SAR imagery is probably no sharper than 1 meter. Resolution varies because the system can cover small areas in greater detail than larger areas. If the imagery meets the wholly plausible NIIRS 6 (1.2–2.5 m) standard for radar, it can detect highway and railway bridges as well as individual vehicles in a known motor pool. And when viewed in pairs or a series, SAR snapshots can help battlefield commanders track the enemy's tanks and trucks.

Real-time radar surveillance is a demanding business both in space and on the ground. For all-weather coverage the United States has relied largely on a pair of radar-imaging Lacrosse satellites, similar in principle to the Magellan space probe sent to map Venus. The space shuttle placed Lacrosse-1 in orbit in 1988, but expendable Titan missiles serviced later missions. Each satellite costs about a billion dollars, weighs roughly 15 tons, and lasts five to ten years. A pair of wing-like solar panels stretching nearly 50 meters from end to end powers a parabolic antenna 10 meters across, which generates 1,500 radar pulses a second. Although a single antenna both sends and receives, the millions of radar echoes it picks up are useless without a high-capacity network of relay satellites and ground stations to deliver the data, and a bank of high-speed computers to reconstruct the terrain.

According to John Pike, *Aviation Week and Space Technology,* and other informed sources, two Lacrosse radar satellites complement two or three KH-12 satellites carrying visible and infrared imaging systems. Although each high-resolution spy satellite can visit trouble spots like Iraq at least twice a day, during the Gulf War images no older than three hours didn't satisfy field commanders, who complained of limited "dwell time." Another gripe was a lack of wide-area coverage. Although money and physics will forever frustrate generals eager to view the entire battle zone in real time all the time, the NRO responded in 1995 with plans for a new 20-ton, $1.5 billion satellite. According to the *Los Angeles Times*, the new satellite would expand coverage of individual scenes from 100 to 800 square miles without any practical loss in resolution. Known as the 8X or the Enhanced Imaging System—the NRO delights in changing code names once the media picks them up—a higher orbit would increase dwell time from five minutes to half an hour. Early in the new millennium the 8X was on duty orbiting the earth 9.7 times a day at an altitude between 2,690 and 3,131 kilometers. As a map on the Heavens-Above Web site confirms, the satellite now known to astronomers as USA 144 had a ground track conveniently close to Iraq and Israel (fig. 2.8).

There's a lot more up there than four or five imaging satellites. Some overhead assets are signals intelligence (SIGINT) satellites, designed to pinpoint defensive radar for the military or monitor communications traffic for the National Security Agency, an NRO cousin with electronic ears in space linked to code-crunching computers outside Washington. Others are measurement and signature intelligence (MASINT) satellites, assigned to track missiles, provide real-time battlefield support, or detect nuclear explosions in space or at the surface. Customized sensors include electro-optical instruments called bhangmeters, which can detect the diagnostic double burst of X rays produced by a nuclear blast, and infrared scanners that can spot and follow a missile's superheated exhaust gasses—missile-interception weaponry like Ronald Reagan's proposed Strategic Defense Initiative, also known as Star Wars and reincarnated as the Bush Missile Shield, rely heavily on thermal tracking.

As with imaging satellites, altitude reflects mission. A low earth orbit (LEO), defined by an altitude typically below 1,000 kilometers, is useful for electronic eavesdropping and thermal detection, whereas a highly elliptical orbit (HEO), in which a satellite's altitude might

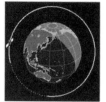 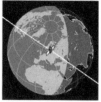

View from above orbital plane   View from above satellite

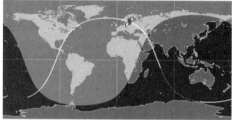
Ground track

FIGURE 2.8 According to the Heavens-Above Web site, the satellite USA 144, aka the 8X image intelligence satellite and the Enhanced Imaging System, passed over the Persian Gulf shortly after 10 P.M. Greenwich time on May 15, 2000.

vary from a few hundred to more than 20,000 km, can be tailored to wide-area or close-look surveillance. By contrast, a geostationary orbit, in which a satellite hovers above the same spot on the equator at an altitude of 36,000 km (22,300 mi.), is useful for satellite telecommunications as well as continuous wide-area surveillance of weather, enemy communications, missile launches, and nuclear explosions. Because an orbit can serve multiple purposes, dual missions are common, especially when satellites work in carefully choreographed groups called constellations. For example, the Air Force's two dozen Navstar satellites, circling at 11,000 km in medium earth orbits (MEO) to support global positioning system (GPS) navigation, also carry nuclear detonation sensors.

Although all satellites require ground support, terrestrial monitors are a key component of nuclear detection. Aside from suspicious activity at the surface, underground tests easily escape satellite detection. Even so, all but the smallest, most carefully hidden atomic explosions produce seismic waves somewhat similar to the tremors of a small earthquake. These waves travel great distances, beyond the borders of a nation violating its test-ban commitment, to register on the worldwide network of seismographs with which scientists estimate the location, depth, and strength of the thousands of earthquakes that occur each year. Fortunately for world peace, seismologists have learned to distinguish nuclear explosions from natural seismic events.

However sharp our eyes and ears, seeing and hearing are vulnerable to false assumptions. On August 16, 1997, for instance, a small earthquake in the Kara Sea, roughly 100 kilometers from a Russian test site, created diplomatic as well as seismic waves. Although seismologists at Columbia University's Lamont-Doherty Earth Observatory knew otherwise, the CIA promptly claimed that the Soviets had detonated an atomic blast. As the story unfolded, it became clear that the CIA had ignored seismic evidence from Russia and Sweden, gotten the location wrong, and incorrectly linked the seismic shock to the test site. Instead of admitting its error, the agency insisted the temblor was an "ambiguous event," which fostered fears in Congress that test-ban treaties could not be verified.

More worrisome is the accidental bombing of the Chinese embassy in Belgrade in May 1999. Shortly after the incident, which killed three Chinese and triggered anti-American riots in Beijing, an acquaintance far more up on remote sensing than I insisted that the bombing was not an accident at all but a way of showing the Chinese how quickly "we" can retaliate should "they" get out of line. Turns out, we're not so arrogant: in a rush to identify bombing targets, CIA analysts who knew the address of a Yugoslavian weapons depot had assumed that house numbers in Belgrade were as orderly as those in Washington and picked the wrong building. Until Congress figures a way to repeal human error, no surveillance network is foolproof.

+ + +

It's clear that satellite surveillance will play a key role in the diplomatic and military history of the twenty-first century. Less certain is whether overhead assets—ours and others'—will prove a hero, a villain, or a bit of both. Essential for treaty verification, satellite imagery is vulnerable to misinterpretation and clever deception. Valuable for threat assessment, and thus vital to national defense and global stability, high-resolution imagery could also trigger an impulsive invasion or a preemptive first strike. As an instrument for identifying human rights abuses and rallying world opinion behind international peacekeeping missions, satellites can foster the New World Order, or merely encourage ineffectual military meddling in Third World countries. As part of an antimissile system like George W. Bush's Missile Shield, satellite sensors might repel or obviate an ICBM offensive, or help venal politicians waste billions of dollars on naïve electronics easily fooled by a salvo of decoys. Equally plausible is the role of high-tech victim: predictably circling in a celestial shooting gallery, intelligence satellites are eminently vulnerable to antisatellite missiles or ground-based lasers that could fry their optics. And that's just for high rollers like the United States, NATO, Russia, and China. Commercial remote sensing beckons still greater ambiguity by suggesting a free market in which high-resolution imagery is readily available to impulsive autocrats, religious zealots, and ethnic purists. Well-intended unilateral efforts to limit access to intelligence-quality photos seem doomed. When sellers outside our borders abound, the "shutter control" favored by the American military merely puts American firms at a competitive disadvantage. International regulation would require that all countries consent to a complex set of rules and sanctions, which is hardly likely according to John Pike, who worries that "every bad guy in the world is going to be buying these pictures."

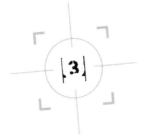

Eyes on the Farm

Agriculture was an early beneficiary of overhead surveillance. During the 1920s and 1930s civil engineers using aerial surveys designed highways, irrigation works, and electric power systems that promoted commercial farming in the semiarid West and brought modern conveniences to rural areas throughout the nation. As an instrument for measuring fields, identifying crops, mapping soils, and monitoring plant growth and pests, air photos also had a direct influence on farming. This role matured rapidly in the 1990s, with the integration of remote sensing, global positioning, and computer modeling. Successful in helping farmers increase production while conserving water and fertilizer, geospatial technology raised privacy concerns when police began using aerial imagery to locate marijuana "grows" and illegally watered lawns.

As independent producers in a free-market system, commercial farmers can be their own worst enemies. The obvious way to increase income is to boost production, but when neighbors harbor similar ambitions, prices can plummet if supply overwhelms demand. This axiom of agricultural economics was seldom more apparent than in the early 1930s, at the onset of the Great Depression, when many farmers found their new tractors more a burden than a blessing. And because falling prices affect all farmers, not just those who mechanize and expand, the Roosevelt administration knew that helping farmers meant stabilizing prices by either buying up the surplus or cutting back production. However implemented, acreage control was crucial: purchasing and storing a year's excess production might avoid excessively high prices if the following year's harvest was poor, but guaranteed prices could easily stimulate further production.

Committed to Secretary of Agriculture Henry Wallace's vision of an "ever-normal granary," New Deal activists took several years to find an effective, constitutionally acceptable way to convince farmers to cut back. The Agricultural Adjustment Act of 1933 focused on production control. Farmers agreed to limit production in exchange for a guaranteed price, partly funded through a tax on processors. To promote "economic democracy" as well as control costs, Wallace turned day-to-day administration of the program over to an innovative hierarchy of state, county, and local committees of farmers.

After the Supreme Court struck down the processing tax in early 1936, emphasis shifted to soil conservation. The Soil Conservation Act of 1936 encouraged soil-building practices and diverted production from soil-depleting crops, which accounted for much of the oversupply. Instead of signing contracts to limit production, farmers applied for payments under whatever programs they were eligible. Production controls returned in 1938, after bumper crops in 1937 proved that conservation incentives alone could not raise farm income. For corn, wheat, cotton, and other critical commercial crops, the Agricultural Adjustment Act of 1938 established national quotas based on past consumption, likely exports, and a reserve for emergencies and prorated these quotas back to the states, counties, and individual farms. Farmers could receive conservation payments only if they accepted voluntary acreage allotments based on the national quota, the local allocation, and their own average yield over the previous ten years.

The need to check compliance and ensure fairness created a measurement crisis. As Howard Tolley, chief of the Agricultural Adjustment Administration (AAA), noted in a 1937 radio interview, "before we can make any payment, we have to find out what each man applying has done to earn it. That means many millions of fields have to be measured." While ground surveys with a surveyor's chain or measuring wheel were slow and costly, air photos promised efficiency and accuracy, especially for irregularly shaped fields. Experiments in a handful of counties in 1935 suggested that a trained worker could approximate ground-survey estimates to within a percent or two by tracing field boundaries on an enlarged print with a carefully calibrated planimeter.

What started as an experimental program quickly became the preferred way to check compliance. In 1937 the AAA planned to map more than 759,000 square miles of farmland in all parts of the country (fig. 3.1). Although existing imagery from commercial firms and other government agencies accounted for slightly over half of the coverage, the remainder was new photography. As AAA technical advisor Harry Tubis told the American Society of Photogrammetry, "the simultaneous photographing of 375,000 square miles of our country, employing 36 photographic crews, is in itself a milestone in the development of aerial mapping."

Coordination and training required a bureaucracy within a bureaucracy. Committees within each of the AAA's five regions prepared guidelines for state and county operations. Each state office had an aerial mapping section that inspected all photos, ordered whatever reflights were needed to plug gaps or correct defects, compiled index maps relating the boundaries of each photo to roads and other local landmarks, and estimated scale from available ground control. In addition, the state sections trained the farm checkers hired by the county committees, spotchecked the checkers' work, and coordinated the ordering of enlarged prints from the Department of Agriculture's new photo labs in Salt Lake City and Washington, D.C. Although the USDA had established a standard scale of 1:20,000 for photo negatives, county offices typically worked with prints enlarged to 1:7,920, the scale at which one inch repre-

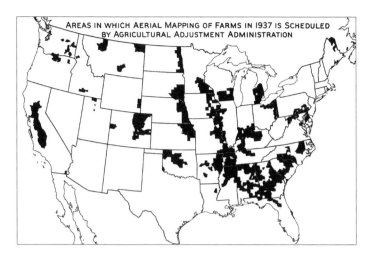

FIGURE 3.1 The AAA's crash program to check compliance focused on the Mississippi Valley, the Southeast, and California's Central Valley. From Harry Tubis, "Aerial Photography Maps Our Farmlands: The Program of the Agricultural Adjustment Administration," *Photogrammetric Engineering* 3 (April-May-June 1937): 22.

sents one-eighth of a mile and a square inch contains ten acres. Enlargement reduced errors likely to arise from imprecise tracing of field boundaries and allowed more detailed annotations for land-use planning.

A potentially important source of error was tilt, the angular deviation of the optic axis of the camera from a plumb line. If tilt is zero, the ground nadir (the point directly below the camera) appears at the exact center of the photo, and if the land is flat and horizontal, scale is constant across the photo. If the photo is tilted, scale generally decreases from the ground nadir toward the photo center and beyond. To correct the problem, engineers in the state office estimated the tilt of each print and supplied "tilt charts" with which planimeter operators in county offices could make an appropriate correction. Although the charts covered tilt as great as 7 degrees, deviations this large were rarely acceptable.

A more troublesome source of error was elevation difference, also called relief. Higher areas in a vertical air photo (typically tableland) appear slightly larger and more detailed than lower areas (bottomland), which are farther from the camera. To compensate, officials in the state office divided photos with significant relief into two or more zones, marked zone boundaries on the print, and provided a scale or correction factor for each zone. Because a single set of zones would simplify calculations at the county level, the state photogrammetrist usually considered both relief and tilt in delineating zones of equal scale.

Where field and pasture boundaries were well established and easily identified, a set of air photos, zone boundaries, and correction tables could serve for a decade or more. Because the photography was intended for measurement, not crop identification, farm checkers would visit each farm applying for payment under the current year's program. Typical of the AAA's insistence on openness and cooperation was the rule that "in no event should the farm checker proceed in the determination of performance on a farm without the knowledge and consent of the owner or operator." The checker would walk the farm with the operator, identify crops, and verify conservation practices. If a farmer disagreed with an evaluation, his objection was noted in the checker's report.

Checkers took the photographs out to the farm for comparison and annotation. Field procedures called for clipping the photo to a rigid board with a smooth surface, protecting it with a waterproof cover, marking farm and field boundaries with a sharp pencil, and identifying farms with a number in red and fields with a letter in blue. Some counties cared for their photos by tracing permanent field boundaries and other features onto a farm map used for measurement and field checking. Farm maps based on ground surveys (quite crude in the AAA's early years) were used in counties without aerial photography. But the government's program of aerial survey was so effective that when the United States entered World War II in late 1941, the USDA had acquired coverage for over 90 percent of the country's agricultural land.

Thanks to the conservation program and its local committees, aerial photography spurred numerous improvements in drainage, plowing patterns, and pasturing. Overhead images presented dramatic evidence of the need for drainage tiles or contour plowing. And an overview of wells, ponds, and catchment basins of streams often suggested a more efficient allocation of livestock. By the late

1940s, conservation improvements had reconfigured more and more field boundaries, and the USDA was dutifully rephotographing the nation's principal farming areas. To stimulate discussion with local committeemen and conservation experts, the Production and Marketing Administration (the AAA's successor in the 1950s) offered farmers enlarged photographs of their land.

Air photos also provided the cartographic foundation for a national soil survey. As with many of its programs, the USDA maps soils county by county. The soil scientist assigned to a county might spend several years in the field assigning categories and delineating boundaries. A typical soil extends several feet below the surface and consists of a series of layers called horizons. Classification is based largely on the thickness, appearance, and physical properties of these layers, which affect fertility as well as suitability for septic tanks, foundations, and other uses. To see what's below the surface, the soil scientist digs small pits or drills narrow holes with an auger. Air photos expedite fieldwork by suggesting good places to sample, by showing moisture or vegetation differences that mirror soil boundaries, and by providing a convenient base for taking field notes and sketching boundaries, which a cartographic technician in the state office transfers to the finished map. Air photos play an even more valuable role in the published report, for which soil boundaries and category symbols are registered onto a visually subdued aerial photographic base map (fig. 3.2). In addition to saving time and controlling cost, the photos show field boundaries and

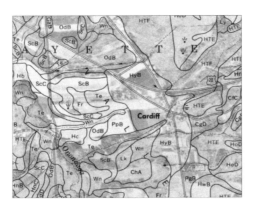

FIGURE 3.2 Portion of a typical soil survey map showing soil boundaries and category abbreviations superimposed on a photomap base. The hamlet is Cardiff, New York, made famous by the Cardiff Giant hoax of 1869. Excerpt from sheet 50 in Frank Z. Hutton, Jr., and C. Erwin Rice, *Soil Survey of Onondaga County, New York* (Washington, D.C.: Soil Conservation Service, 1977).

other features that make the soil information easily accessible to farmers, planners, and civil engineers.

+ + +

Longstanding links between agricultural cartography and military intelligence fostered a fuller use of aerial imagery in farming. The relationship began in the 1920s, when photointerpreters trained by the army during World War I returned to work at agricultural colleges and the USDA and began using air photos for soil mapping, conservation planning, and forest management. When the United States entered World War II, the USDA provided the military with experienced photogrammetrists as well as the services of its two photo labs. When the war ended, many more imagery-savvy GIs returned to jobs in soil science, agronomy, forestry, and farming. Soils experts and agronomists in particular were eager to experiment with a recent military innovation—infrared imagery, designed to detect camouflage—that was promisingly proficient in delineating moist soils and stressed crops.

Before the advent of camouflage-detection film, the opposing army could hide its tanks and supplies under mottled green tarps and buildings painted green to resemble dense shrubbery from the air. As figure 3.3 shows, live vegetation and camouflage reflect the blue, green, and red light similarly. Both appear green in color to the human eye. But extension of the reflectance curves into the near-infrared (*infrared* meaning "beyond red") portion of the spectrum, where wavelengths are a bit longer and beyond human perception, reveals markedly different spectral signatures. In this part of the spectrum, green canvas or netting reflects much less light than does

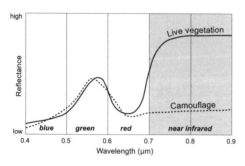

FIGURE 3.3 Spectral signatures of healthy vegetation and camouflage are generally similar for the visible light portion of the electromagnetic spectrum but notably different for near-infrared wavelengths.

leafy vegetation. The vegetation thus produces a much brighter image on the camouflage-detection film, which is sensitive only to infrared light. (Note that near-infrared light, sometimes known as reflected-infrared light, with wavelengths between roughly 0.7 and 1.1 micrometers, is not the same as thermal-infrared (heat) radiation, which has wavelengths between 3 and 20 micrometers.)

In a 1943 magazine ad, Eastman Kodak demonstrated the efficacy of camouflage-detection film with the trio of photographs in figure 3.4. A large factory with a water tower, loading dock, and railway siding can be concealed to look like a wooded area on standard black-and-white film. As the ad explains, infrared film is not so easily fooled. "Natural grass and foliage contain chlorophyll—Nature's coloring matter. Camouflage materials lack this living substance. Chlorophyll reflects invisible infrared light, making the natural areas look light in the picture—almost white. In violent contrast, the 'dead' camouflaged areas show up dark—almost black—in the pictures." Instead of fooling bomber pilots and photointerpreters, ersatz woodland betrayed a promising target.

FIGURE 3.4 With standard black-and-white (panchromatic) film, camouflage netting and artificial trees *(center)* disguise a factory *(left)* that would make an attractive bombing target. Dark, dead-looking tones on an infrared image *(right)* expose the fakery. Reprinted courtesy of Eastman Kodak Company. KODAK is a trademark.

Color-infrared imagery, a related innovation, makes leafy crops and other foliage stand out like the clichéd sore thumb. Conventional color film uses separate emulsions to capture the intensities of reflected blue, green, and red light, and color prints with corresponding dyes produce a realistic composite image. Color-infrared film earns its other name, false-color film, by replacing the blue layer with an emulsion sensitive to infrared radiation and rearranging the bands so that infrared prints as red, red masquerades as green, and green becomes blue. Ignoring the blue band is beneficial:

blue light is readily scattered in the atmosphere and gives normal color air photos their distinctive and disconcerting haze. In color-infrared photos, clear blue water, which absorbs green, red, and infrared light, looks black, and healthy vegetation, which reflects infrared radiation most efficiently, appears bright red.

Intrigued with what air photos might reveal about terrain and hostile forces, military researchers experimented with films and filters. The reflectance characteristics of vegetation and soil were key elements in terrain studies focusing on concealment, enemy food supplies, and the ease with which troops and vehicles could move across open country. Although most of the work was classified, an official from the Navy Photographic Interpretation Center described an agriculturally significant project at the 1953 meeting of the American Society of Photogrammetry. At the request of the National Research Council's Committee on Plant and Crop Ecology, naval researchers had measured reflectance for diseased and healthy cereal crops. A key exhibit compared the spectral signatures of wheat with varying degrees of leaf damage caused by rust spores. As the curves show, severely damaged wheat reflects more reddish radiation and less infrared radiation than less severely affected wheat (fig. 3.5). An accompanying photograph taken with infrared film showed a strong contrast between the dark tones of severely diseased crops and the light tones for fields with relatively little damage.

Overhead imagery would discover a lot more than crop stress. Experiments with photographic and electronic sensing systems led to effective methods for classifying crops, measuring soil moisture, and delineating soils that are chronically dry or chemically deficient.

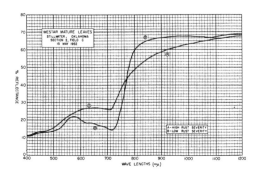

FIGURE 3.5 Military experiment showing different spectral signatures for severely and moderately diseased wheat. From Page E. Truesdell, "Report of Unclassified Military Terrain Studies Section," *Photogrammetric Engineering* 19 (1953): 470.

Multispectral scanners designed to record reflected radiation in comparatively narrow, carefully chosen portions (bands) of the electromagnetic spectrum proved especially useful in identifying fields needing attention. Thermal and microwave radar scanners were equally promising. Although reflected-infrared imagery can point out moisture differences only at the surface, thermal sensors can describe moisture conditions in the soil's upper 10 centimeters. Equally important, synthetic aperture radar (SAR) systems for sensing soil moisture can penetrate the crop canopy as well as provide the all-weather imagery essential for routine, periodic monitoring.

+   +   +

Until very recently commercial satellite imagery has been too coarse for mapping farms at the subfield level, and airplanes and helicopters have carried most of the sensors used in crop management. Color-infrared images from Landsat's multispectral scanners, with resolutions between 79 and 30 meters, provided vivid documentation of changing agricultural land use but did not afford the detail needed to optimize applications of fertilizers, herbicides, and pesticides. The 20-meter resolution of SPOT's multispectral scanner was not noticeably more informative.

Temporal resolution can be more troublesome than spatial resolution. For example, Landsat-7's repeat cycle of sixteen days is much too long for effective crop management even though its scanners were substantially more useful to agriculture than earlier Landsat sensors. The farmer who applies insecticide two weeks late will have already lost the crop.

Even so, medium-resolution satellite imagery has a role in precision agriculture. Research at the USDA's Water Conservation Laboratory in Tucson, Arizona, showed that Landsat-7's Enhanced Thematic Mapper Plus (ETM +) data can be integrated with crop growth models as well as SAR data from current and future earth-imaging satellites. More frequent radar data calibrated to a Landsat vegetation index can partly compensate for the long repeat cycle, and numerical models based on daily meteorological measurements and a detailed soil map can simulate plant growth, soil conditions, and evaporation. Promising developments include expedited processing of Landsat data, growing international interest in radar satellites, improved crop-management models, and inexpensive transmitter-equipped sensors for monitoring conditions in the field. In addition to trouble-shooting maps that highlight areas needing fertilizer, water, or pesticide, process models also generate yield maps and profit maps that help farmers explore diverse planting and harvesting strategies.

The four maps in figure 3.6 demonstrate the scope and scale of subfield analysis. Their focus is a 45-acre field near Salina, Kansas. As the map in the upper left shows, soybean yield in 1999 varied from less than 20 to more than 50 bushels per acre. At the upper right a second map describes the pattern of electrical conductivity (EC), which was measured with a probe-profiler towed behind a tractor. EC, which is negatively correlated with yield, reflects grain size, texture, and other soil properties that affect root development and plant growth. Performance Benchmark Analysis, a software solution that compares yield to the soil's potential productivity, generated the map at the lower left, on which varying shades of gray identify areas performing below their potential. Where the performance level is less than 90 percent, as shown on the map at the lower right, the model estimated the cost of underperformance in dollars per acre. Nitrogen fertilizer and other treatments can significantly increase yield in these areas.

Multiple technologies are typical of precision agriculture, which uses GIS to manage data and support modeling, remote sensing to monitor crops and soil, and GPS to guide sprayers, harvesters, and other machinery. The goal is increased profitability, and the strategy is integration and control. Closely linked sensors and computer models not only advise the farmer but collect data in the field. For example, an automatic variable-rate manure spreader with a GPS, a radio, and its own sensors can send the GIS updated reports on soil conditions and weed density. In addition to directing the application of water and fertilizers, spatial data and process models might eventually control the movement of unmanned agricultural vehicles.

It's clear that "farming soils, not fields," can significantly raise profit margins. Pierre Robert, director of the University of Minnesota's Precision Agriculture Center, compares the increased productivity of information technology to three earlier innovations: the

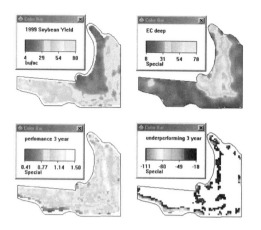

FIGURE 3.6 Subfield maps describing a 45-acre field's 1999 soybean yield *(upper left)*, electrical conductivity *(upper right)*, 1999 performance level *(lower left)*, and the cost of underperformance in dollars per acre *(lower right)*. The original maps are in color. Courtesy of Veris Technologies, a division of GeoProbe Systems.

tractor, agrochemicals, and hybrid crops. Adoption will be slow, he cautions: "It took more than 30 years to see tractors fully utilized." Less clear is the farmer's future role as an independent decision maker and entrepreneur. Crop consultants monitor fields and advise grain farmers on planting, cultivating, and harvesting. In the Midwest, many farmers rely on "custom cutters," who own the combines and hire the crews for efficient, timely harvesting and hauling. As corporate farms grow larger, smaller producers cede greater control to banks, marketing cooperatives, and biotechnology firms hawking genetically engineered seeds. It's hardly surprising that crop consultant Dennis Berglund would wonder whether "the data will [eventually] be worth more than the land."

+   +   +

Adept at monitoring production and controlling weeds, geospatial technology seemed an obvious tool for law enforcement officials eager to find and destroy an outlawed but highly profitable weed, marijuana. But with different goals and ground rules, precision pot busting has yet to replicate the success of precision farming. Effective against greedily indiscreet domestic producers as well as Third World peasants proud of their poppies, remote sensing loses much of its power when savvy cultivators scatter their plants across cornfields and forests. Because cannabis is not markedly different in its spectral signature from most other herbaceous plants, small or sparsely mixed plots are impossible to spot from space and difficult to discern from a plane or helicopter. Even so, authorities are rightly suspicious of cultivated plots in wildlife sanctuaries or on federal land that shouldn't have tended fields with scarecrows.

When imagery won't work, drug enforcement agencies can combine a GIS with an expert system designed to outsmart intrepid growers who circumvent asset forfeiture laws by planting pot in national forests. In the early 1990s geographers Delvin Fung and Roy Welch developed a prototype for Chattahoochee National Forest in northeast Georgia. Intended to predict locations likely to contain pot plantations, their system integrated information about the ecology of cannabis and the behavior of cultivators with data on terrain, roads, streams, and forest cover. Results were impressive: their system not only predicted seventy-five of ninety previously identified marijuana sites but targeted three others for field checking by Forest Service police, who found cannabis planted at all three.

Producers who try to avoid overhead surveillance by moving their plants indoors must be wary of forward-looking infrared (FLIR) sensors, designed originally for nighttime warfare. Windowless hydroponic pot farms replicate natural sunlight with full-spectrum "growlights" that emit both visible and ultraviolet light. In addition to energizing the plants, growlights warm the air, which in turn radiates excess heat through walls and roof. Aimed from ground or air, thermal-infrared imaging systems reveal the excess heat as a bright spot, warmer than its surroundings. That's why thermal imaging is useful for energy-conservation agencies studying heat loss as well as for cops ferreting out clandestine "grow rooms." Because heat from growlights can look a lot like heat from a fireplace or poorly insulated attic (fig. 3.7), thermal infrared alone doesn't justify a search warrant. But when further investigation links an anomalous hot spot to other suspicious activity, few judges will refuse a warrant.

Recent postings to the HempCultivation.Com Security forum reflect growers' concerns about thermal imaging. "James Bong" wondered whether "anyone had detailed information on the use and accuracy of thermal imaging devices on plants of our favour." His query triggered a lengthy reply from "Ganga Warrior," a nongrower (I assume) eager to share experiences and insight.

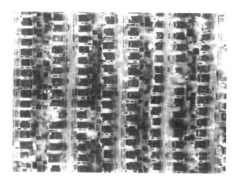

FIGURE 3.7 Thermal-infrared image of a residential neighborhood showing unusual heat loss for a few homes with overly warm interiors or poor insulation. The small bright dots on most roofs pinpoint chimneys. Courtesy of the Rochester Institute of Technology—Center for Imaging Science.

As one of the cops involved in detecting and eradicating dope, I can tell you that thermal imaging is not used much for outdoor grows.

In the early 1990s when peace broke out, the DoD and other agencies with hi-tech equipment that was suddenly under-used came to law enforcement and touted their wares. We tried everything from computer analyzed, false color photography, IR, and thermal imaging. None of them were very effective on outdoor grows.

Thermal imaging is still VERY BIG with indoor grows because of the heat generated by the cultivation.

Infra red (IR) and thermal gear worked best on outdoor grows when used during the heat of the day. The water lines, being cooler than the surrounding terrain and vegetation, show up very well, but there was no way to tell if they were domestic water supply or supporting a grow.

Perhaps the most insidious use of this spy technology was the attempts to use sensitive equipment to detect the minute amounts of radio isotopes in the galvanized chicken wire and the low level electrical impulses put out by batteries in the water timers. These happened at a level much higher than my lowly position so I don't know if they had any success.

Ganga's irreverent remarks are incisive. Drug interdiction is a big business for law enforcement and equipment vendors, and skirmishes over pot are part of the wider war against more potent drugs like cocaine and heroin. Trickle-down from military research encourages integration of diverse monitoring, screening, and tracking technologies, including the unmanned aerial vehicle (UAV) in figure 3.8, equipped with a video camera for "undetected surveil-

FIGURE 3.8 A government grant helped the Pima County, Arizona, Sheriff's Department develop an unmanned aerial vehicle (UAV) equipped with a video camera and transmitter. From Office of National Drug Control Policy, Counterdrug Technology Assessment Center, *Confronting Drug Crime and Abuse with Advanced Technology* (Washington, D.C., 2000), 13.

lance." Even so, police rely mostly on informants, anonymous tips, and the occasional visual survey with a helicopter.

+   +   +

What police consider a clever way to trawl for indoor marijuana farms, civil libertarians see as an assault on the Fourth Amendment, which bans "unreasonable searches and seizures." At issue is a household's expectation of privacy and the extent to which infrared sensors can look through walls and roofs. For the most part, judges have sided with the police. In 1991, for instance, in *United States v. Penny-Feeney*, the Federal District Court in Hawaii held that using a thermal scanner was a reasonable way to check out an anonymous tip and better establish probable cause for a search warrant. Authorized to search Janice Penny-Feeney's home, county police found ten 1,000-watt growlights and 247 marijuana plants. In rejecting Penny-Feeney's petition to suppress the evidence, the court concluded that the sensor sees nothing more than "waste heat"—the legal equivalent of stench from a garbage bag abandoned at the curb for constitutional removal by the sanitation department, snooping neighbors, or enterprising cops. Judges in the Seventh, Eighth, and Eleventh Circuits apparently bought this argument, although additional evidence helped establish probable cause.

Not all jurists agree. In 1998, for instance, in *United States v. Kyllo*, a three-judge panel in the Ninth Circuit Court of Appeals diverged on the obtrusiveness of FLIR scanners. Two of the judges rejected the waste heat excuse and overturned indoor marijuana grower Danny Lee Kyllo's conviction because of a "presumptively unreasonable" warrantless search. Although police had obtained a warrant to search Kyllo's home, their contention of probable cause was based partly on a ground-based FLIR scan. "Even if a thermal imager does not reveal details such as sexual activity in a bedroom," the court's majority declared, "with a basic understanding of the layout of a home, a thermal imager could identify a variety of daily activities conducted in homes across America: use of showers and bathtubs, ovens, washers and dryers, and any other household appliance that emits heat." Not so, their dissenting colleague argued: the scanner merely "measured the heat emanating from and on the outside of a house" and "intruded into nothing." The government requested a rehearing, and sixteen months later the appeals court reversed itself, restored Kyllo's conviction, and pronounced the scanner not "so revealing of intimate details as to raise constitutional concerns."

Kyllo appealed to the Supreme Court, which in mid-2001 ruled in his favor 5 to 4 and sent the case back to the lower court. Writing for the majority, Justice Antonin Scalia declared the search unconstitutional because warrantless use of "sense-enhancing technology . . . not in general public use" violates a citizen's reasonable expectation of privacy. Justice John Paul Stevens, who wrote the dissent, countered that the case involved "nothing more than drawing inferences from off-the-wall surveillance, rather than any 'through-the-wall' surveillance." To bolster his position that the images were not intrusive, Stevens referred to the contested thermal image of a hot spot near the middle of Kyllo's roof, above the growlights (fig. 3.9), and observed that "the device could not . . . and did not . . . enable its user to identify either the lady of the house, the rug on the vestibule floor, or anything else inside the house." The staunchly conservative Scalia was not impressed: "In the home," he maintained, "*all* details are intimate details, because the entire area is held safe from prying government eyes." Civil libertarians were elated at the decision, which one observer called "surprisingly broad." Even so, the ruling neither prohibits police from obtaining a warrant for FLIR surveillance nor addresses the possible proliferation of cheap thermal scanners sold over the Internet or at the local RadioShack.

FLIR image of the
Kyllo home

FLIR image with
interpretative
annotations

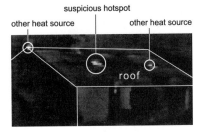

FIGURE 3.9 Thermal image revealing a plausibly incriminating hot spot near center of Danny Lee Kyllo's roof. The unannotated image was a defense (government) exhibit in *Kyllo v. United States*.

Curiously, the *Kyllo* decision diverged from the high court's traditional conservative-liberal split. Justice Scalia's libertarian opinion drew support from not only his equally conservative colleague Clarence Thomas but also the moderately liberal justices Steven Breyer, Ruth Bader Ginsburg, and David Souter. By contrast, Justice Stevens, arguably the court's most liberal member, was joined in his dissent by the conservative Chief Justice William Rehnquist as well as the moderate conservatives Anthony Kennedy and Sandra Day O'Connor. Because views on abortion and school vouchers seem to have little bearing on a justice's attitude toward remote surveillance, it might be difficult without pointed probing to predict a future nominee's stance on surveillance technology.

Equally problematic is the legality of overhead imagery. The last time the high court ruled on aerial photography was 1986, in *Dow Chemical Company v. United States*. Dow had charged the Environmental Protection Agency with committing an unconstitutional warrantless search by hiring an aerial survey firm to inspect its 2,000-acre Midland, Michigan, plant for violations of the Clean Air Act. In ruling in favor of the EPA, the court found that taking pictures of the sprawling plant from navigable air space was perfectly constitutional because the exposed pipes, machinery, and other facilities were more like an "open field" than the "curtilage" (courtyard or fenced-in yard) around a home. Simply put, Dow had no right to expect privacy because "the intimate activities associated with family privacy and the home and its curtilage simply do not reach the outdoor areas or spaces between structures and buildings of a manufacturing plant." Moreover, Congress had authorized the EPA to search for violations.

Although the 1986 court waffled on whether use of "highly sophisticated surveillance equipment not generally available to the public, such as satellite technology, might be constitutionally proscribed absent a warrant," the "open field" argument in *Dow* clearly justifies satellite surveillance like the use of SPOT imagery to ferret out illegal irrigation. But don't expect agreement from ardent privacy advocates. After the *Wall Street Journal* reported the irate protest of a farmer charged with irrigating 39 acres without a permit, libertarians were outraged that the Arizona Department of Water Resources would compare satellite imagery with computer records of irrigation permits. In a Libertarian Party denunciation of "Big Brother" and "spy satellites," national chairman Steve Dasbach complained that "high-tech military equipment . . . once used against foreign enemies [was] now being used against American citizens on a routine basis." Arizona was not the only snoop: several states were smoking out pot with thermal scanners, Georgia planned a satellite search for illegal timber cutting, and North Carolina tax assessors were scrutinizing high-resolution satellite imagery for improved properties, soon to be hit with bigger tax bills. I doubt that many libertarians, most of whom abhor taxes, government fees, and government in general, appreciated DWR director Rita Pearson's retort that satellite imagery helped her department cut back its enforcement staff.

If the Supreme Court sets further limits on overhead surveillance, the justices might well draw a line at comparatively obtrusive sensors like active microwave radar. Scanners that generate their own radiation are arguably more menacing than devices that detect waste heat or reflected sunlight. Even so, distinctions between residential and nonresidential buildings seem crucial, as the high court implied in *Dow* and reaffirmed in *Kyllo*. Although the judiciary might distinguish between acceptable government enforcement and excessively intrusive private uses, high-definition satellite imagery will probably pass the "open field" test. Space Imaging no doubt thought so when its house magazine—in an article titled "Sizing Up the Competition: Earth Information Takes Commercial Intelligence to a New Level"—encouraged overhead surveillance of plants, warehouses, and shipping sites by investors and competitors. Increased availability of overhead imagery over the Internet from Space Imaging and at Web sites like terraserver.com defuse much of the high court's unease in *Dow* (and more recently in *Kyllo*) about "highly sophisticated surveillance equipment not generally available to the public."

Tinder, Technology, and Tactics

Wildland fire tops the list of hazards I might have included in my book *Cartographies of Danger* but didn't. It seemed reasonable to omit a hazard so dependent on the current day's weather and the number of years since the last big burn. Besides, I needed to limit the book's scope and length, and no one seemed to be making hazard-zone maps similar to those for tornadoes, floods, and earthquakes. No doubt my oversight also reflects an East Coast bias. Had I lived in Montana or New Mexico, fuel maps and fire atlases would have received the respect accorded groundwater models and toxic plumes.

In June 2000, when I began to plan this chapter, wildland fire was all over the news. That May the Cerro Grande Fire—a "prescribed fire" that got out of control—destroyed more than two hundred homes at Los Alamos, New Mexico, and toward the end of summer, when I started to write, wildfires were on front pages and nightly newscasts almost daily. By then I had a fuller appreciation of the scope and devastation of western fires. For added understanding, a vacation that July included visits to Los Alamos as well as Mesa Verde National Park, in southwest Colorado, where a 1996 fire burned nearly 5,000 acres.

What I saw in Los Alamos spoke of individual tragedy and collective naïveté. Residential development had expanded upslope into scenic woodlands, which rebelled by bringing the fire into backyards and vestibules. Remnant chimneys and burned-out cars testified to the hazards of the "urban-wildland interface"—no less a high-risk zone than a floodplain or shoreline. In some neighborhoods haphazard devastation suggested an evil lottery had targeted one in every four or five homes at random. Chain-link fencing encircled rubble-strewn lots with charred foundations, while life went on next door with few apparent scars. Was it luck, fire-retardant roofs, or successfully deployed lawn sprinklers?

By contrast, Mesa Verde was recovering nicely. Although broad expanses of blackened, leafless trees marked the extent of the 1996 fire, wildflowers and small shrubs had aggressively colonized the recently exposed soil. And in a stroke of serendipity, the fire that burned off the piñon and juniper had uncovered hundreds of Ancestral Puebloan artifacts, eagerly mapped and cataloged by National Park Service archaeologists. But because the spring and early summer had been exceptionally hot and dry, areas that escaped the conflagration four years earlier were now in far greater danger. As my wife, Marge, and I toured the cliff dwellings and agricultural ruins, we saw fire crews busily clearing brush and cutting trees near some of the park's more important sites. Reducing the amount of readily combustible fuel can slow a fire and make it easier to extinguish. We also noticed helicopters carrying huge buckets of water to new fires—small ones, thankfully—started by lightning the night before. Suppressing minor fires while they're still small demands vigilance and luck. Two weeks later the luck ran out: a wildfire triggered by lightning burned a third of the park and sent archaeologists out to the fire lines, to protect previously unknown sites from firefighters' axes.

+   +   +

To understand geographic technology's role in coping with wildland fire requires an appreciation of the factors that influence ignition and spread. Foremost, of course, is fuel. Anything living or dead that will burn, release heat, and feed the fire is considered fuel, whether it's in the ground, on the surface, or in the air. Ground fuel, which includes tree roots, buried wood fragments, and decaying organic matter in the soil, can spread a fire to nearby surface material and is difficult to extinguish when dry. Fire experts distinguish surface fuels like short trees and shrubs, fallen branches and needles, and other surface litter from aerial fuels like the trunks and branches of trees taller than about five feet and their leaves, needles, mosses, and lichens. In this upper layer, a crown fire fed by conifer needles and other fine materials can spread rapidly from tree to tree. Crown fire is less likely with deciduous leaves, which burn readily only when very dry. Even so, dead leaves add to surface fuel, which is more prone than aerial fuel to ignition from a lightning strike or careless camper.

In addition to differentiating among ground, surface, and aerial layers, a fuel inventory must account for the mass, moisture content, and flammability of forest fuels. Compiling a comprehensive fuel inventory can be a complex, costly, and slow process if the forester conscientiously selects sample plots, identifies species, weighs woody material, and carefully estimates its moisture content. Lacking time and budget, forest managers typically delineate boundaries between forest types and assign each zone to a risk category.

During the 1930s, when fuel mapping was relatively new, the U.S. Forest Service focused on identifying intuitively hazardous areas and training field workers to be consistent when using photographic keys to assign categories. Figure 4.1, from a mid-1930s fire control manual, describes the symbols employed in mapping fuels on 15 million acres of federal and privately owned "cooperative" lands. Designed to focus attention on high-risk areas, the map uses contrasting colors to describe the likely rate of spread of a small fire and parallel lines to show the number of firefighters required to put it out. Superposition of what might have been two separate maps recognizes the enhanced hazardousness of areas in which ground cover, steep slope, or other factors would slow construction of a fire control line.

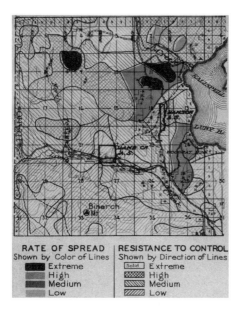

RATE OF SPREAD
Shown by Color of Lines
- Extreme
- High
- Medium
- Low

RESISTANCE TO CONTROL
Shown by Direction of Lines
- Solid Extreme
- High
- Medium
- Low

FIGURE 4.1 Portion of 1930s-era field fuel map for Kaniksu National Forest in Idaho. The numbered squares are square-mile Public Land Survey "sections," each of which contains 640 acres. The original map is in color, with black, red, green, and yellow-orange representing extreme, high, medium, and low risk of spread, respectively. From L. G. Hornby, *Forest Fire Control Planning in the Northern Rocky Mountain Region* (n.p.: U.S. Forest Service, Rocky Mountain Forest and Range Experiment Station, 1936), 51.

Contemporary fuel maps reflect wildfire-specific vegetation categories based on carefully controlled laboratory "burns." Experimental measurements of the temperatures attained and amounts of heat released by various kinds of plant matter, alive or dead, provide a foundation for numerical "fuel models" that encompass an array of flammability factors, including the presence of oils, resins, and other highly volatile compounds and the effect of diameter on the rate of drying. Grasses, dry needles, and lichens, for example, can gain or lose most of their moisture in little more than an hour, whereas logs over three inches in diameter might take more than a month to respond to exceptionally wet or dry conditions. Fuel models also consider the amount of time (sometimes measured in centuries) over which surface fuels have accumulated as well as the spacing of burnable material, which might be packed too tightly to produce flames that could spread the fire to nearby litter.

Besides helping the Forest Service prioritize brush removal and other fuel reduction efforts, fuel models are used to assess current fire danger and predict the behavior of actual fires. The National Fire Danger Rating System integrates mid-afternoon weather data with a set of twenty fuel models and calculates a series of indexes representing the potential for surface fires over large areas for a twenty-four-hour period. Each fuel model is named for a type of vegetation; examples include mature chaparral, heavy logging slash, and "closed short-needle conifer (heavily dead)." Depending on the values of the Burning Index, the Spread Component, and the Energy Release Component, officials make an educated guess about whether the next twenty-four-hour period will be a normal day, a fire day, a large-fire day, or a multiple-fire day—categories that reflect demands on personnel and equipment. (As interpreted verbally for campers and passing motorists, a simpler system rates fire danger as low, moderate, high, very high, or extreme.) To predict the behavior of an observed wildfire, the Forest Service relies on a set of thirteen fuel models, which provide coefficients for mathematic equations describing a wildland fire's intensity, flame height, and rate of spread.

Fire-spread models must also consider wind and slope. On a flat, uniform plane a fire expands outward in a circular pattern, as hot gasses from the burning material and thermal radiation from the flames dry out and ignite areas just beyond the flames. Add a steady wind and the circle becomes an ellipse. As figure 4.2 illustrates, even a light, 1-mile-per-hour wind produces a noticeably elongated fire perimeter, and higher wind speeds yield more pronounced elongations.

Not all fires behave like a simple contagious process. As the left-hand diagram in figure 4.3 illustrates, a fire advances more readily in the direction of a steady wind, which pushes superheated gasses forward near the surface, tilts the flames outward over the fuelbed

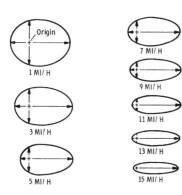

FIGURE 4.2 Fire-spread models reflecting the elongating effect of wind speed on fire shape. Redrawn from figure IV-4 in Richard C. Rothermel, *How to Predict the Spread and Intensity of Forest and Range Fires*, USDA Forest Service Research Paper Series, no. INT-143 (June 1983), 63.

Wind-driven fire

Upslope fire

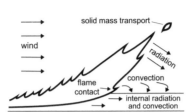

FIGURE 4.3 Factors in the spread of wind-driven and upslope fires. Redrawn from figures 2 and 3 in Richard C. Rothermel, *A Mathematical Model for Predicting Fire Spread in Wildland Fuels,* USDA Forest Service Research Paper Series, no. INT-115 (January 1972), 5.

directly downwind, and carries airborne ashes, embers, and other "firebrands" well beyond the flame front. Because fires can reinforce the wind by creating turbulent vortexes called fire swirls, shaggy fragments of burning bark can stay lit and aloft long enough to "spot" a new fire more than a mile away. Ellipse-based models designed to represent surface fires cannot predict the growth of large fires, which typically expand by spotting and "breakouts." Richard Rothermel, the Forest Service engineer who pioneered fire-spread models, was well aware of their limitations. In describing the hazards of fighting large fires in very dry conditions, he observed that "the fires literally make their own weather, mostly rain and winds called downbursts. That's how they suddenly blow up."

Terrain further distorts the fire perimeter. As the right-hand diagram in figure 4.3 shows, a fire will advance more rapidly upslope as radiation from its flames attacks the full height of trees and shrubs immediately ahead. Because an upslope fire can "explode" and outrun a firefighter, few places are as dangerous as a ridge crest directly ahead of an advancing wildfire. Narrow canyons are equally hazardous because radiation and embers from one side of the corridor will expedite the fire's advance along the other side and vice versa. And a gully aligned with a strong upslope wind is akin to a flamethrower pointed at anyone on the ridge above.

Given a fuel model, a wind velocity, and a uniform slope, a fire-spread model can calculate the likely perimeter of a small fire several hours after ignition. Because the fire perimeter becomes less regular as the fire grows and conditions change, later estimates are usually based on two or more "projection points" (numbered 1 and 2 in fig. 4.4) representing the forward advance of the fire front. The strongest wind seldom coincides with the steepest slope, so fire control officers treat these effects as vectors and use their resultant to project the fire perimeter forward.

In the 1970s, when fire-spread models were still largely experimental, fire officials typically estimated the expected rate of advance for different areas with graphs or a computer program and transferred the results by hand to a topographic map. Predictions were handicapped by coarse boundaries between fuel types and irregular terrain. As Rothermel cautioned, accuracy depends "upon the skill and knowledge of the user and the degree of uniformity or lack of uniformity of the fuels and environmental conditions." Even so, the fire-spread model is a useful tool for making an educated guess about the most effective way to suppress the fire and protect firefighters.

+ + +

Computers handle most of the map work nowadays. Although fire-spread models are still valuable in suggesting worst-case scenarios and planning the prescribed fires used to reduce the fuel hazard, emphasis is on real-time management, which includes tracking the

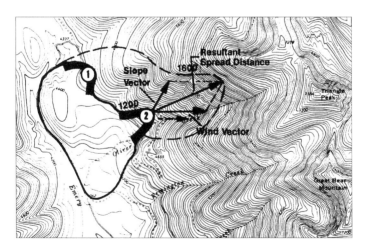

FIGURE 4.4 Projected advance of a fire perimeter based on projection points (1 and 2) and separate distance vectors for slope and wind. Compiled from figures IV-7 and IV-9 in Rothermel, *How to Predict the Spread and Intensity of Forest and Range Fires*, 78–79.

fire front and monitoring the current positions of fire crews and fire engines. Geographic information systems excel at integrating information from such diverse sources as weather maps, satellite imagery of vegetation, maps of fire trails and other local roads, and digital terrain models, which describe accessibility and provide slope data for fire-spread models. Heavily dependent upon telecommunications, modern fire fighting is much like modern warfare as fire crews with two-way radios and handheld GPS receivers update their map coordinates, a helicopter pilot with a GPS unit traces the fire perimeter, and the "fire boss" coordinates communications with community fire departments.

Because GIS software is not nearly as user-friendly as its vendors would have us believe, each of California's twelve Interagency Incident Management Teams includes a GIS specialist. For small fires, a notebook computer with a portable printer and plotter are usually adequate. For larger fires, the team's technical specialist will request a mobile GIS unit and additional "tech specs" to operate the multiple workstations and large-format plotter. California's pyro-cartographers depend on a variety of data, including satellite images, digital elevation models (DEMs) used in fire-spread modeling, "fire-ground intelligence" based on GPS readings, thermal-infrared imagery from Forest Service aircraft, and a set of nine CDs containing topographic data for the entire state. Terrain maps extracted from the CDs form the basis for an Incident Action Plan and illustrate the morning and evening briefings at which fire officials evaluate the situation and assign personnel.

Keeping track of who's where and doing what is especially challenging for large fires like the October 1995 Vision Wildfire, which consumed 11,410 acres at Point Reyes National Seashore in northern California. The fire burned for twelve days, and at its height engaged 196 fire engines, 27 bulldozers, 7 helicopters, 7 tanker aircraft, and 2,146 personnel, including 74 fire crews. Emergency officials assembled a team of four GIS specialists within twelve hours of the fire's ignition. A day and a half later they had set up a fully equipped GIS lab—quite an accomplishment in 1995, when mobile labs were just a cool idea—and were busy plotting updated maps of the fire perimeter and damaged structures. When the fire was out and most firefighters had returned home, GIS experts were busy mapping the impact on wildlife and rare plants and assessing the fire's response to fuel type, slope, and wind. Retrospective modeling is useful for training firefighters as well as making residents and local officials aware of the dangers of living in the urban-wildland interface. According to Sarah Allen, the "tech spec" who directed the effort, GIS maps are "great tools for public relations" as well as priceless data for local communities eagerly seeking federal disaster relief funds.

+ + +

Because large fires are difficult to manage, much less model, efficient suppression depends on early detection. The principal line of defense used to be a network of fire towers and other fixed, ground-based lookouts staffed by observers who scanned the horizon with binoculars for plumes of smoke and reported in by telephone or shortwave radio. Back in the 1980s GIS helped fire officials reconfigure lookout networks to provide maximal coverage with a minimal number of conveniently accessible observation posts. Nowadays geographic technology provides additional sets of eyes, watching from space and often able to see at night and through clouds or smoke. Detecting forest fires is another role for weather and earth observation satellites already in orbit for other purposes.

A satellite's ability to detect fire depends on its orbit and sensors. Two complementary orbits seem especially appropriate. As configured for the National Oceanic and Atmospheric Administration's Polar-orbiting Operational Environmental Satellites (POES), a low-altitude near-polar orbit provides comparatively detailed, moderately frequent worldwide coverage. As shown schematically in figure 4.5, an orbital plane inclined slightly to the earth's axis moves steadily westward as the satellite circles the globe fourteen times a day in a mildly elliptical orbit. A sensor scanning a ground swath 1,700 miles (2,800 km) wide from an altitude of approximately 520 miles (840 km) provides twice-a-day coverage for mid-latitude locations, and a pair of satellites in orbital planes 90 degrees apart reduces the revisit time to six hours. By contrast, one of NOAA's Geostationary Environmental Observations Satellites (GOES), parked 22,300 miles (35,900 km) above a fixed point on the equator, stares downward at slightly less than half the globe but reports in with a new image every fifteen minutes. This markedly higher orbit limits GOES imagery to a resolution of 2.5 miles (4.0 km) directly over the equator, with progressively larger pixels and fuzzier images with increased distance from the nadir point. The comparatively lower POES orbit compensates for less frequent coverage with sharper images based on 0.7-mile (1.1 km) pixels along the nadir track.

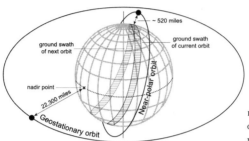

FIGURE 4.5 Comparison of geostationary and near-polar orbits.

POES's prowess as a fire lookout depends on thermal-infrared imagery from its AVHRR (Advanced Very High Resolution Radiometer) scanner. Designed for meteorological, oceanographic, and hydrologic studies, the AVHRR is a multispectral scanner similar in principle to the Landsat and SPOT sensors (see chapter 2), which capture separate images for carefully selected portions of the electromagnetic spectrum. Among the AVHRR sensor's five bands sampling visible, near-infrared, and thermal energy, band 3 (3.55–3.93 μm) is especially suited for detecting wildland fires, in which temperatures range from 500 to 1,000 kelvins (400–1,300°F). Although insensitive to temperature differences within a large fire—pixels reach their maximum brilliance at a "saturation temperature" of only 320 kelvins (120°F)—the band-3 sensor is ideal for detecting hot spots as small as an acre and thus valuable for recognizing fires shortly after ignition, when suppression is most effective.

Imagery and maps for the June 1996 Millers Reach, Alaska, wildfire demonstrate the accuracy and feasibility of near-real-time monitoring. Like most high-latitude locations, Alaska enjoys comparatively frequent AVHRR coverage because POES orbits converge near the poles such that ground swaths often overlap. In reconstructing the fire's evolution, NOAA researchers further reduced the six-hour revisit time by integrating band-3 AVHRR imagery with data from ground-based Doppler radar, the GOES Imager, and polar-orbiting Defense Meteorological Space Program satellites. Despite the geostationary satellite's comparatively coarse resolution (8 km at 60°N), the fire's approximate location was readily apparent shortly after ignition on a GOES thermal-infrared image similar in sensitivity to AVHRR band 3. GIS analysts now knew where to look for further growth and were able to track the expanding burn zone by superimposing various images onto a map with a 1-kilometer grid (fig. 4.6). By 0533 LST (local sun time) on the morning of June 5, the fire was centered on three large hot spots surrounding Big Lake. Arrows on the map show where strong winds carried the fire to the west and south over the next eight hours and more than doubled the number of pixels and acres affected. In addition to successfully tracking a fire from space, imagery used in the study was available within forty minutes after ground stations received the data.

Geostationary satellites offer a promising platform for detecting and monitoring wildfire in real time. Although a GIS can enhance reliability by integrating imagery from complementary sources, noth-

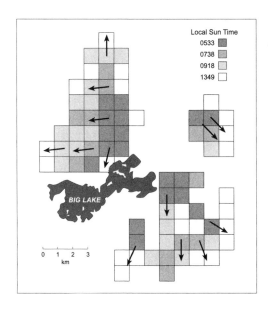

FIGURE 4.6 Fire-affected pixels identified on imagery from NOAA and Defense Department meteorological satellites describe the advance of the Millers Reach fire, in south-central Alaska, over an eight-hour period on June 5, 1996. Movement of the fire front can be tracked outward from the darker pixels, which burned earlier than the lighter pixels. Redrawn from figure 12 in Gary L. Hufford and others, "Use of Real-time Multi-satellite and Radar Data to Support Forest Fire Management," *Weather and Forecasting* 13 (1998): 603.

ing beats the timeliness of a fresh snapshot every fifteen minutes. Sharpen its spatial resolution with a specially designed sensor, and a GOES successor might even pinpoint small fires. That substantial improvements are possible is implied by an international commission's call for "an ultimate detection time of 5 minutes, a repeat time of 15 minutes, spatial resolution of 250 meters, and real time transmission to local ground stations." Although this commission, the Committee on Earth Observation Satellites (CEOS), touts the advantages of a sharper, more timely God's-eye view, cloud cover would still be troublesome because thermal signals cannot penetrate thick clouds.

Equally intriguing is the Langley Research Center's proposal for FireSat, a research satellite that would monitor biomass burning worldwide. Proposed in the mid-1990s as a five-year scientific mission but still on NASA drawing boards a half-decade later, FireSat could customize satellite sensors for mapping forest and grassland fires as part of a larger concern with the effect of wildland fire on global climate. A comparatively inexpensive "small sat" would carry a 23-kilogram (50 lb.) scanner in a circular near-polar orbit at an altitude of 520 miles (830 km) and provide four daily passes over northern forests, once-a-day surveillance of middle latitudes, and every-other-day coverage of the tropics. The scanner's seven bands would be optimized for detecting smoke and particulate matter; mapping active fires and recently burned areas; detecting cirrus clouds, vegetation moisture, and fire scars; and measuring surface temperature. Other significant departures from the AVHRR and GOES Imager sensors would include 263-meter (860 ft.) resolution along the nadir track and a saturation temperature of 1,000 kelvins (1,300°F), about the maximum temperature found in wildland fires.

Exceptionally sharp imagery and a high saturation temperature are essential for accurate maps of fire intensity. Although a 500 × 500-meter (1,600 × 1,600 ft) pixel would provide valuable portraits of burn scars and fire fronts, a markedly sharper image, with smaller pixels, is needed because hot air and turbulence would surely blur the picture. According to research on the uncertainty of pixel measurements for active fires, half the radiation measured for a 263-meter pixel might originate outside its nominal ground spot. FireSat compensates with comparatively fine resolution from which deblurring software could recover acceptably sharp images. Even so, this extraordinary spatial detail would be wasted if the sensor allowed the pixels to saturate well below the fire's maximum intensity.

FireSat's originator and most persistent advocate is Joel Levine, a senior research scientist in Langley's atmospheric sciences division and a key collaborator in national and international studies of climate change. Levine's chief concern is the contribution of fires—both deliberate and natural—to increased amounts of carbon dioxide and particulate matter as well as the threats of higher temperatures, deforestation, and ozone depletion. FireSat is partly an effort to document the extent of burning, which affects about 1 percent of the earth's surface in a typical year and is much more widespread than once believed. Humans set most of the fires, principally to clear land for agriculture.

A contributor to the CEOS proposal, Levine is skeptical of the commission's call for a fifteen-minute repeat time, which, he notes, would require "a constellation of sixty satellites." Nevertheless, an image frequency of "an hour or less" seems both feasible and essential, especially if NASA needs support from the Forest Service and the Federal Emergency Management Agency (FEMA), which see fire suppression and timely evacuation as more compelling than scientific studies of global warming. To win their backing, Levine's team would place a second FireSat instrument on a geostationary satellite.

For now, FireSat is only a feasibility study, not an authorized mission. NASA management wants to be sure the mission can deliver everything that its champions claim. To address these concerns, Langley researchers have been busy calibrating existing satellite imagery with temperatures measured on the ground in prescribed fires and working with aerospace-electronics firms to develop a lightweight, long-lived thermal sensor. Levine is philosophical about the delay: if FireSat becomes the world's first dedicated fire-monitoring satellite, it must rely heavily on new, innovative technology designed by the space agency's commercial partners.

+   +   +

Useful in detecting wildland fires, satellite imagery can also identify areas requiring intensified brush cutting as well as additional firefighters and equipment. AVHRR imagery is one of several ingredients in the national fire potential index (FPI) map developed by fire researchers at the Forest Service in collaboration with remote sensing experts at the U.S. Geological Survey. Another element is GOES, which helps update the map by relaying data from remote automatic weather stations. The FPI map treats the country as a system of one-square-kilometer pixels and integrates twenty-four fuel models with detailed maps of land cover and vegetation. Each cell's fire potential index is updated daily by moisture data from nearby weather stations and a "relative greenness" index, which relates the cell's current vegetative vigor to its recorded highest and lowest values. As in camouflage detection and precision agriculture, near-infrared energy trumps visible green as a decisive indicator of vegetative greenness.

No state has exploited relative greenness and other FPI concepts more effectively than Oklahoma, which integrates satellite imagery with a state-of-art, real-time weather data network, the Oklahoma Mesonet. Intended for state and local fire officials, the Oklahoma Fire Danger Model produces five separate maps eleven times a day. Figure 4.7 illustrates two of them: the spread component map, which estimates the advance in feet per minute of a wind-driven fire front, and the ignition component map, which estimates the likelihood that firebrands will cause a fire requiring suppression. The model produced these maps for 2 P.M., April 8, 1999, when a prairie fire might have moved more than 120 feet a minute across the state's Panhandle and moderate winds increased the threat of firebrands. Red and orange symbols (dark areas on my black-and-white version) underscore the danger toward the southeast of wildland fire to semiarid grasslands, which can dry out and burn much more rapidly than woodlands.

If you're curious about fire potential or active fires in your area or a favorite vacation spot, visit GeoMAC Wildland Fire Support Web site (geomac.cr.usgs.gov). In cooperation with the Forest Service, NOAA, the National Interagency Fire Center, and several other federal agencies, the Geological Survey provides a cartographic overview of active fires and generalized fire perimeters as well as detailed

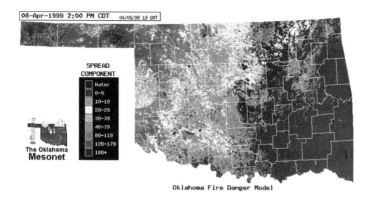

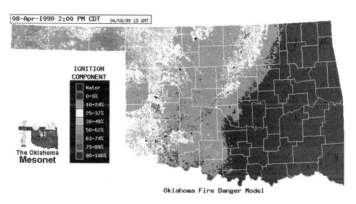

FIGURE 4.7 Spread component *(top)* and ignition component *(bottom)* maps from the Oklahoma Fire Danger Model for 2 P.M., April 8, 1999. From the Oklahoma State University, *AgWeather*, http://agweather.mesonet.ou.edu/models/fire/.

maps of individual fires based on imagery from space, fixed-wing aircraft, and ground locations. The Web is a convenient medium for distributing maps and other data to fire managers and field personnel, and there's no reason to restrict public access. We're paying for the information, after all, and it might make some of us more wary of the hazards of the urban-wildland interface.

+ + +

Although satellite imagery and geographic information systems are immensely useful for detecting and suppressing wildland fire, a focus on putting out forest fires would be myopic. In the American

West and other dry areas, accumulated biomass that is not removed by prescribed fire or brush clearance will eventually burn—this year, next year, or maybe not for a hundred years. And fuel spared this year can support a bigger fire next year. Clear-cutting is an even worse solution: the tall trees removed for lumber are far more resistant to fire than the shrubs that follow or the litter left behind. And the more environmentally correct but costly practice of clearing brush by hand is practicable only for maintaining firebreaks or protecting specific structures. Equally inefficient are grazing animals—don't look for Smokey Goat to displace Smokey Bear as a fire-prevention icon. At the end of the day, the most dependable strategy is prescribed fire.

As a national policy, fighting fire with fire opens broader roles for GIS and FireSat. Improved modeling will help fire managers plan and control prescribed fires, while enhanced monitoring affords more timely warnings and evacuations as well as more reliable decisions about when and where to let a natural fire burn. Equally important are the roles of geographic surveillance in promoting awareness of urban-wildland hazards, in establishing and enforcing realistic zoning regulations and building codes, and in setting insurance rates that allow for occasional miscalculations like the Cerro Grande Fire. Ultimately unavoidable, wildland fire is less dangerous when fewer people live in the woods.

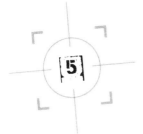

Weather Eyes

More pervasive than wildfire, destructive storms demand vigilant surveillance by a vast network of probes and sensors. Networking is crucial: although each measurement of temperature, pressure, wind, and humidity could contain an important clue to the atmosphere's next assault, individual numbers mean little without a cartographic framework that reveals meaningful patterns and ominous trends. In much the same way that weather maps helped nineteenth-century scientists unravel the nature of storms, computer models and satellite images are the mind and eyes of modern forecasting.

Geographic surveillance must accommodate a variety of meteorological scales, from hurricanes to lightning strikes. Large rotating storms known as hurricanes or northeasters were hard to predict when the forecaster's only eyes were land-based instruments. Still not fully understood, big storms are easy to spot from space, and surveillance technologies save thousands of lives by giving local officials ample warning of when and where evacuation is warranted. Less predictable are smaller, more localized phenomena like tornadoes and wind shear, for which the foreseeable future is a matter of minutes, not days. Lightning strikes are too small for exact forecasting, but precise after-the-fact maps warn of possible wildfires and identify substandard power lines.

+   +   +

In *Isaac's Storm,* Erik Larson recounts the tragedy of Galveston, Texas, where more than six thousand people died in September 1900, when an unanticipated hurricane of exceptional strength surged in from the Gulf of Mexico. Larson's hero is Isaac Cline, head of the local U.S. Weather Bureau office. Personal diaries and agency archives tell a tale of foolish pride and bureaucratic paralysis. Forecasters in Washington controlled weather data for the entire country and issued all warnings of serious storms. Although pressure and wind data cabled from Cuba suggested an enormous low-pressure system had entered the Gulf, the Central Office questioned the competence of Cuban meteorologists and ignored their observations. Although wary of rising seas along the Galveston waterfront, a dutiful Isaac Cline echoed the bureau's denials and discouraged evacuation until disaster was imminent. Cline lost his wife, his house, and his trust in Chief Willis Moore, who in the investigation that followed flagrantly distorted the Weather Bureau's actions.

However inexcusable, Washington's incompetence reflects a fatal mix of feeble theory and sparse data. Reconstructing an offshore storm from coastal observations is tricky when forecasters must infer the presence of a low-pressure center from a handful of barometric readings along its periphery. Washington officials knew that a tropical storm had passed over Cuba but assumed it was headed north across Florida like previous hurricanes. Turn-of-the-century forecasters believed that a storm center would tend to follow the generalized path of similar storms traversing the same area in the same direction at the same time of year. Dismissing wind and waves along Florida's west coast as mere "offshoots" of a hurricane more likely to threaten the Carolinas, Weather Bureau officials had no idea that Isaac's storm had slipped into the Gulf of Mexico's warm waters, where it gathered strength for a fatal assault on Galveston. According to their maps, hurricanes that crossed Cuba didn't track northwestward into the Gulf.

Another recent bestseller paints a radically different portrait of meteorological monitoring. The antagonist in Sebastian Junger's *The Perfect Storm* is an intense, well-developed "nor'easter" that formed in the North Atlantic in late October 1991. Its victims were the six crewmembers of the *Andrea Gail,* a commercial fishing boat based in Gloucester, Massachusetts. This time the National Weather Service (so renamed in 1967 and now a part of NOAA) got it right. Satellite imagery had tracked a severe cold front that moved off the northeast coast of the United States. Forecasters knew rogue waves were likely when an enormous, near-record high-pressure system behind the front led to an intense extratropical cyclone along the front, east of Nova Scotia.

Extratropical cyclones are large circular storms that, unlike hurricanes, form outside the Tropics. This one, later named the Halloween Storm of 1991, expanded rapidly and even absorbed Hurricane Grace, which had recently crossed the Atlantic from the west coast of Africa, where most hurricanes originate. On October 27, forecasters chose the rarely used, deliberately ominous adjective *perfect* to warn mariners that a "dangerous storm" would form over the North Atlantic within thirty-six hours. Rather than ride out the tempest on the Grand Banks with its sister vessel, the *Hannah Borden,* the *Andrea Gail* headed back to Gloucester with its catch. The following evening the ship lost radio contact after reporting 30-foot waves and 80-knot winds 180 miles east of Canada's Sable Island, near the center of the intensifying storm.

The story doesn't stop there. On October 29, the system moved southward and westward to flood homes and devastate beaches from Canada to North Carolina. Although the storm's lingering presence caused hundreds of millions of dollars of damage along the East Coast, low-lying areas were evacuated and fewer than a half-dozen coastal residents drowned. As the satellite image in figure 5.1 demonstrates, the storm's size sent a stern warning to anyone tempted to stay behind.

FIGURE 5.1 GOES-7 visible image for 1 P.M. EST, November 1, 1991, showing the extent of the Halloween Storm with an unnamed hurricane at its center. From NOAA's satellite art gallery, NOAA Climatic Data Center, http://www.ncdc.noaa.gov/ol/satellite/satelliteseye/.

This November 1 satellite photo also captured the meteorological oddity of a storm within a storm. A huge counterclockwise swirl describes a vast but dying storm still raining on the Northeast. At its center is a much smaller system with a clearly marked eye. At 6 P.M. New York time, when this snapshot was taken, the storm entered the record books as the Unnamed Hurricane of 1991 after winds near its center reached 65 knots—just above the official threshold of 64 knots (74 MPH). American and Canadian forecast officials agreed not to name the hurricane because the storm had already inflicted its greatest damage and would soon break up. Naming it, they figured, would create needless confusion about storm forecasts and evacuation warnings.

Hurricane Andrew, which rampaged across south Florida the following August, underscores the importance of timely evacuation. As hurricanes go, Andrew was not huge, but it was intense, with sustained winds up to 145 miles per hour. Meteorologists at the National Hurricane Center in Miami had tracked the storm's development since August 14, when satellite imagery revealed an anomaly in the trade winds west of Africa. Andrew reached hurricane strength on August 22 and collided with the Florida coast two days later. As the satellite image in figure 5.2 confirms, NOAA forecasters had ample time to initiate a timely evacuation of more than a million people in Dade, Broward, and Palm Beach Counties. Although the storm caused $25 billion in damage and entered the record books as the nation's costliest natural disaster, only sixty-five people perished, directly or indirectly. Had Andrew taken Floridians by surprise, it would have eclipsed the 1900 Galveston storm as the country's deadliest natural disaster.

According to Louis Uccellini, who directs NOAA's National Centers for Environmental Prediction, "though our forecasts were quite good for [the 1991 Halloween storm], we can do even better today." Participating in a June 2000 roundtable discussion of the Perfect Storm—the movie version of Junger's book was setting box-office records at the time—Uccellini cited marked improvements in storm prediction, which can look ahead as far as ninety-six hours. Although computers run the forecast models, he noted, satellites provide 85 percent of the data. What's more, by signaling the earli-

FIGURE 5.2 NOAA-12 satellite image showing Hurricane Andrew approaching the Florida coast on August 23, 1992. Original color image from NOAA's Historical Significant Events Imagery database, NOAA Climatic Data Center, http://www.ncdc.noaa.gov/pub/data/images/hurr-andrew-19920823-n12rgb.jpg.

est appearance of a tropical disturbance, satellite imagery helps weather officials plan aircraft reconnaissance missions, which yield valuable supplementary data.

NOAA relies on two types of weather satellites: geostationary platforms like GOES-7, which captured the black-and-white snapshot of the Perfect Storm in figure 5.1, and polar-orbiting satellites (POES) like NOAA-12, which caught Hurricane Andrew advancing on Miami in figure 5.2. Although a POES platform circling the globe at an altitude of only 520 miles can take sharper pictures, its GOES counterpart 22,300 miles above the equator allows almost constant coverage of a far broader area and more vigilant surveillance of a storm's position, size, and intensity. Robert Sheets, who directed the National Hurricane Center (now the Tropical Prediction Center) in the late 1980s, was a strong advocate of the GOES gaze. In a recapitulation of the NHC's trials and triumphs, Sheets opined that "if there was a choice of only one observing tool for use in meeting [the center's responsibilities, I] would clearly choose the geostationary satellite."

NOAA's strategy calls for a minimum of two active satellites, a GOES-East mission covering eastern North America, the North Atlantic, and Latin America, and a GOES-West mission for the western United States and the central and eastern Pacific. Although GOES satellites have a design life of five years, contingency plans now include a standby satellite that can be repositioned if one of the assigned satellites fails. Relocation is possible because of tiny rockets called thrusters, also used to fine-tune the orbits of geostationary satellites. When the GOES-West satellite died in 1989, thrusters moved its eastern counterpart westward to a temporary GOES-Prime position covering the continental United States. After NOAA borrowed and repositioned a spare European satellite to cover the North Atlantic, thrusters moved the remaining GOES platform farther westward for fuller surveillance of the Pacific. International cooperation is well established in meteorology, and United States geostationary satellites fill two of five niches in a worldwide weather surveillance system that includes satellites maintained by India, Japan, and a consortium of European weather services.

Think of GOES as a platform, not a camera. The more recent GOES vehicles—GOES-12, launched in July 2001, is similar in design to GOES-8, placed in orbit in 1995—carry a variety of sensors as well as transponders for relaying measurements from a network of buoys and other surface stations. The most prominent sensor is the GOES Imager, with resolutions between 1 and 8 kilometers at ground nadir. Designed for frequent wide-area monitoring of atmospheric moisture, including cloud movement that reflects wind currents, the Imager can be reprogrammed to provide timely snapshots of small areas threatened by severe weather. Complementing the Imager is the independently programmed GOES Sounder, which uses a single visible band and eighteen thermal-infrared channels to monitor water vapor, temperature, cloud height, and other atmospheric conditions. Soundings taken hourly inform numerical forecast models about the current state of atmospheric moisture.

Three imaging schedules provide focused coverage of severe weather as well as preferential treatment of American territory. In routine operation, the GOES-East Imager produces CONUS (conterminous United States) and southern-hemisphere portraits at a half-hour interval and full-disk images (fig. 5.3) at a three-hour interval. Fifteen minutes after each CONUS portrait, a wider shot of the northern hemisphere reduces the interval for the forty-eight contiguous states to a quarter-hour. When unstable air within the focal region calls for more frequent monitoring, the Imager's "rapid-

FIGURE 5.3 Full-disk image from GOES-7 showing Hurricane Andrew advancing on the Louisiana coast at 2 P.M. CDT, August 25, 1992. Original color image from NASA Goddard Space Flight Center, Hurricane Andrew photo gallery, http://rsd. gsfc.nasa .gov/rsd/images/andrew .html.

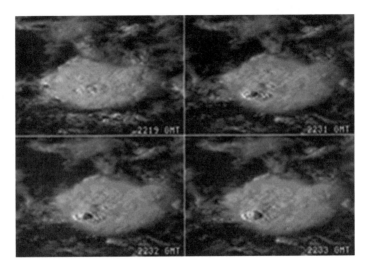

FIGURE 5.4 GOES-8 captured the "explosive" development of a thunderstorm over Tulsa, Oklahoma, during a twelve-minute period on July 20, 1994. The final three of these super-rapid-scan images are only a minute apart. From NOAA National Weather Service, Southern Region Headquarters, NOAA Satellite Tutor, http://www.srh.noaa.gov/maf/SatTutor/cira_rscan.gif.

scan" mode can reduce the CONUS interval to 7.5 minutes on average at the expense of southern-hemisphere coverage, which is cut back to only once an hour. Even more timely monitoring is possible with the Imager's "super-rapid-scan" mode, which abandons the southern hemisphere entirely, except for the three-hour full-disk image, in order to cut delivery time to five minutes or less for a specially designated SRSO sector. Able to scan a 1,000 kilometer by 1,000 kilometer sector at a one-minute interval, SRSO mode can capture volatile phenomena like the upwardly protruding thunderstorm in figure 5.4.

Under a similar plan, the GOES-West satellite provides routine quarter-hour coverage of Alaska, Hawaii, and the western states with alternating northern-hemisphere and PACUS (Pacific United States) scenes a half-hour apart. And like its eastern counterpart, the west-coast Imager produces southern-hemisphere and full-disk portraits at half-hour and three-hour intervals, respectively, in routine mode and reduces its southern-hemisphere coverage during rapid-scan and super-rapid-scan operation. By contrast, the east and west GOES Sounders each have only a single operation plan, which supplements hourly CONUS scenes with less frequent oceanic views and focused soundings of smaller areas such as the Gulf of Mexico and the eastern Caribbean Sea.

The success of satellite-assisted hurricane forecasting lies partly in the storms' size and slow movement, and partly in the sophisticated computer models used to predict storm position. According to a Tropical Prediction Center study of cyclone tracking in the Atlantic Basin between 1970 and 1998, errors for twenty-four-hour forecasts declined by 1.0 percent per year. Marginally more impressive are decreases of 1.7 and 1.9 percent per year, respectively, for forty-eight- and seventy-two-hour forecasts. After adjusting for erratic storm behavior, which increases the difficulty of making reliable predictions, NOAA researchers observed distinctly higher rates of improvement—2.1, 3.1, and 3.5 percent per year—for the period 1994 to 1998. Even so, average errors of 84, 151, and 221 nautical miles impede efforts to forecast a storm's position one, two, or three days hence.

Officials of NOAA and the Federal Emergency Management Agency (FEMA) find this uncertainty troubling, and because many places cannot be evacuated in twenty-four hours, they "overwarn" just to be safe. As a result, hurricane warnings issued a day ahead of landfall typically affect 400 nautical miles of coastline even though a tropical hurricane might leave a trail of destruction only a hundred miles wide. Until track forecasts as well as predictions of storm size and intensity are more precise and reliable, overwarning is preferable to underwarning, even though public confidence might slip when a storm dies or goes elsewhere after triggering a massive evacuation.

+   +   +

To further improve prediction, hurricane forecasters are using radar to probe the internal structure of storms. Unlike passive sensors, which depend on reflected sunlight or thermal radiation, radar is an active sensor, which measures the reflectance of its own electromagnetic pulses and can penetrate clouds as well as operate at night. Especially revealing are three-dimensional radar images from the Tropical Rainfall Measuring Mission (TRMM), a research satellite launched in November 1997 by NASA and its Japanese counterpart. In a circular orbit inclined 35 degrees to the plane of the equator, the TRMM meanders across the tropics at an altitude of 215 miles (350 km). Among the TRMM Observatory's five principal instruments is the world's first space-borne precipitation radar, which supplied revealing views of hurricanes Bonnie, Irene, and Floyd well before they were within range of land-based radar. In addition to describing a storm's horizontal extent, imagery like the overhead snapshot of Hurricane Floyd in figure 5.5 affords a revealing three-dimensional view of vertical variations in precipitation. Moisture profiles akin to those in the upper- and lower-right of this NASA publicity photo allow more reliable estimates of the amount and intensity of precipitation as well as a more accurate picture of the center of circulation within a storm. Rainfall within a hurricane is not uniform, and a sense of where precipitation is likely to be most intense can be useful in last-minute emergency preparations.

Land-based radar is also helpful in tracking hurricanes and warning residents of locally intense rain and tornado-force winds. During Hurricane Andrew in 1992, officials at the National Hurricane Center gained a firsthand appreciation of weather radar when the storm center passed just south of their building. The radar map (fig. 5.6) shows a donut-shaped zone of intense precipitation and turbulence a few miles southeast of the NHC. Although damage is usually more pronounced in coastal locations subject to flooding, Andrew's winds caused considerable destruction inland, thanks to a disastrous combination of inadequate building codes, lower land values, and cheap construction. As indicated by a circular band of high winds just outside the storm's eye, devastation was well underway south of Miami. This map gained a prominent place in weather lore when, minutes later, winds gusting to 164 miles per hour swept the radar antenna off the hurricane center's roof.

The following April, Miami received a newer, more efficient weather radar system, NEXRAD (*Nex*t Generation *Rada*r). A cornerstone of the National Weather Service's modernization plan, NEXRAD replaced radar systems implemented in 1957 and 1974. The new system employed Doppler radar, able to measure the direction and speed of wind as well as differentiate among dust, rain, sleet, snow, and hail. Under a program started in 1988, the weather

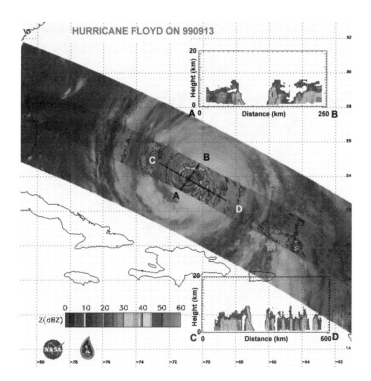

FIGURE 5.5 TRMM precipitation radar provides cross-sectional views (profiles A-B and C-D) showing horizontal and vertical variations in precipitable moisture. This NASA publicity image captures Hurricane Floyd approximately 500 miles east-southeast of Florida on September 13, 1999. Original color image from NASA's TRMM Web site, http://trmm.gsfc.nasa.gov/data/2000_data/HurrFloyd1_md.gif.

service had begun replacing older radars with NEXRAD and reducing the number of forecast offices. Because of the new radar's greater range, NOAA reasoned, fewer offices were required. Cities that lost forecast offices complained, but overall NOAA's new radar network is a marked improvement over its predecessor. NEXRAD is credited with increasing the average warning time for tornadoes from five to twelve minutes, which can be especially significant if trailer parks have shelters, homes have basements or "saferooms," and residents receive timely warnings through NOAA's Weather Radio network, a local broadcaster, or a community siren system. Early detection is useless when likely victims can't take shelter.

Although NEXRAD is a clear success, estimates of the average

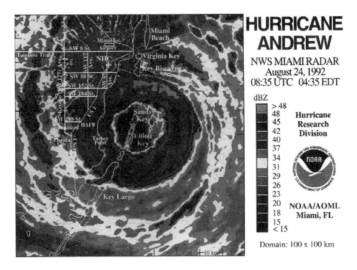

FIGURE 5.6 Radar image of Hurricane Andrew showing a ring of high winds and intense precipitation a few miles away from the National Hurricane Center (marked with a plus-sign labeled "NHC"). Minutes later, winds toppled the center's radar antenna. Original color image from NOAA Office of Public and Constituent Affairs, image gallery, http://www.publicaffairs.noaa.gov/photos/1992andy.JPG.

lead time for tornado warnings vary widely. For example, a 1996 cover story in *Time* magazine reported a NEXRAD-induced improvement from three to eight minutes, while a 1999 proposal by a subcommittee of the National Science and Technology Council (a consortium of federal agencies) claimed an end-of-decade average between fifteen and twenty minutes. By contrast, a NOAA case study published on the Internet by the National Partnership for Reinventing Government (another interagency collaboration) described a steady advance in lead times from nine minutes in 1995, to ten, eleven, and twelve minutes in 1996, 1997, and 1998, while an August 2000 NOAA "goals statement" reported an eleven-minute average—"nearly triple the three minute lead time of 1977"—and suggested that a fifteen-minute average is possible by 2005. These and other claims accord with the Knight Ridder Tribune newspaper syndicate's WeatherQuiz, which credits NEXRAD with extending the run-and-hide time from five to twelve minutes.

Despite the confusion, lead time improved significantly during the late 1990s, when computers learned to interpret radar maps more vigilantly and reliably than the best human forecasters. Because radar maps are visually complex, timely warnings depend upon computer algorithms that recognize suspicious signatures like the "hook echoes" of precipitation being drawn into a rotating storm. Developed, tested, and fine-tuned by teams of forecasters and research meteorologists, computer-assisted surveillance also contributed to an impressive increase in the lead time for flash-flood warnings—up from eight minutes in 1987 to fifty-one minutes in 2000.

Averages hide a fascinating diversity in the storms that generate tornadoes and the protection afforded by weather radar. At a June 1999 hearing before the House Committee on Science, Dennis McCarthy, head meteorologist at the National Weather Service Office in Norman, Oklahoma, described the carnage of May 3, six weeks previous, when more than seventy tornadoes ravaged Tornado Alley and killed forty-two people in Oklahoma alone. Without NEXRAD, the death toll would have been much higher. McCarthy and his colleagues sent out 116 county-level warnings for tornadoes and severe thunderstorms, and "false alarm rates were much lower than average." Lead times for these tornado warnings averaged eighteen minutes—a bit better than the Norman office's average of fifteen minutes, which in turn is higher than the national average of eleven minutes. Still more impressive is the thirty-two-minute lead time for warnings issued for southern Oklahoma City and nearby Moore. McCarty attributed the accuracy and timeliness of these predictions to NEXRAD: "Doppler radars are like CAT scans of the atmosphere enabling forecasters to see wind fields within thunderstorms. These thunderstorm wind fields are precursors to tornadoes."

How thoroughly NEXRAD can scrutinize a thunderstorm depends partly on the storm's size and proximity. Most of Oklahoma's May 3 twisters were byproducts of tall, wide, and relatively long-lived thunderstorms called supercells, which NEXRAD can detect as far as 180 kilometers (112 mi.) away. Proximity is important because the earth's surface curves away from the straight-line radar beam, which at some distance becomes simply too high to detect low-altitude phenomena like tornadoes. Increased distance from the antenna also widens the radar beam, which lowers resolution, an important ingredient of weather radar's CAT-scan effect.

NEXRAD is thus less likely to pick up an offshoot tornado's distinctive low-altitude hook echo beyond about 100 kilometers (62mi.). With a generous thirty-two minutes of lead time, residents of Moore and southern Oklahoma City were no doubt fortunate that Norman's NEXRAD site and highly experienced team of tornado forecasters were less than 10 miles away.

Hall County, Georgia, a rural area about 40 miles northeast of Atlanta, was not so lucky. Around 6:25 A.M. on March 20, 1998, a tornado struck without warning and killed eleven people. According to Don Burgess, who trains forecasters in the use of NEXRAD, the thunderstorm that spawned the March 20 tornado was a mini-supercell, so named because of its narrower diameter and lower top. Harder to detect than its bigger relatives, a mini-supercell is "a challenging event for a forecaster." Hall County had an added disadvantage: the nearest NEXRAD station was more than 80 miles away, in Peachtree City. According to a 1995 National Research Council study, NEXRAD cannot detect mini-supercells more than about 100 kilometers (62 mi.) away, and cannot reliably discern related hook echoes beyond 70 kilometers (43 mi.).

Sensitivity to storm size and height account for minor cracks in the NEXRAD safety net. According to the National Research Council, NOAA's new radar provides much better surveillance for supercells than for mini-supercells. Moreover, the 15 percent of the contiguous states that is either hidden by terrain or outside the maximum detection range of at least one antenna lies largely in remote, sparsely populated parts of the mountainous West. By contrast, the map of mini-supercell detection (fig. 5.7) reveals substantial gaps in the East and Midwest. NRC calculations suggest that 68 percent of the contiguous territory is comparatively vulnerable to this less common and (fortunately) less destructive brand of severe weather. Even so, a small storm cell can be tracked once it's detected, and NOAA encourages forecasters to examine images from neighboring radars as well as consult with neighboring forecasters and officials at the Storm Prediction Center, in Norman, Oklahoma. Because NEXRAD is truly a network, telecommunications can stitch together the islands of surveillance on its mini-supercell coverage map.

+ + +

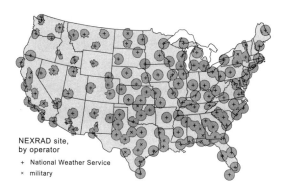

NEXRAD site,
by operator
+ National Weather Service
× military

FIGURE 5.7 Areas of mini-supercell detection by NOAA's NEXRAD network. Compiled from figure 2-4a in NEXRAD Panel, National Weather Service Modernization Committee, Commission on Engineering and Technical Systems, National Research Council, *Toward a New National Weather Service: Assessment of NEXRAD Coverage and Associated Weather Services* (Washington, D.C.: National Academy Press, 1995), 23.

A comparatively sparse network monitors vertical variation in wind currents. At thirty locations in sixteen states, largely in the Central Plains, experimental Doppler radars called Wind Profilers gaze upward past the top of the troposphere to sample wind speed and direction every six minutes. To provide a vertically dense sample, the profilers alternate between two slightly overlapping altitude ranges: "low mode" takes readings every 250 meters (820 ft.) between 500 and 9,250 meters above the ground, while "high mode" provides similar scrutiny between altitudes of 7,500 and 16,250 meters—that's 10 miles up, well beyond the ceiling for most commercial aircraft.

For each profiler location, a time-section chart such as in figure 5.8 describes temporal variation in wind velocity. The vertical axis of the chart shows elevation above sea level and the horizontal axis represents time, with the most recent readings on the left and the oldest on the right. For each height and time a thin, arrow-like staff portrays wind direction and its barbs signify wind speed, with each full barb representing 10 meters per second and a short barb indicating 5 meters per second. Thus, a line with two full barbs and one short barb represents a wind speed of 25 meters per second. Orientation is shown relative to north (toward the top) and east (to the right), as on a map, and by convention wind staffs point downwind.

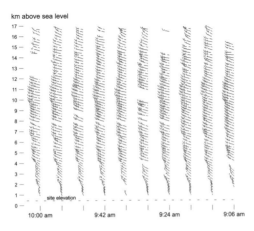

km above sea level

FIGURE 5.8 Time-section chart describing winds above Fairbury, Nebraska, at 9:06 to 10:00 A.M. CT on November 27, 2000. The dashed line represents Wind Profiler site's elevation of 0.43 kilometers above sea level. Redrawn from an illustration published on the NOAA Profiler Network Web site, http://www .profiler.noaa.gov/jsp/ index.jsp.

Thus at 10 A.M. the wind at 12 kilometers above sea level was flowing from slightly north of west at approximately 30 meters per second, while at elevations below 2 kilometers air was moving much more slowly from the north-northwest. Missing data vectors indicate measurements rejected by the Wind Profiler's quality control processor because of weak signals or radio interference.

Time-section charts offer a variety of insights. Strong, consistent, high-altitude winds like those between elevations of 10 and 11 kilometers in figure 5.4 suggest the presence of the jet stream—a fact readily confirmed by making a high-altitude winds map from available profiler reports. Winds maps are important to forecasters because upper-air currents can steer storms and deliver precipitation. In addition, radical differences in wind direction within a row or column might indicate turbulence or wind shear, which pilots try to avoid. Wind Profiler data also provide a reality check on numerical models. This complementarity was especially valuable in the hours preceding the May 3, 1999, tornado outbreak, for which NEXRAD provided valuable local warnings once a profiler in Tucumcari, New Mexico, alerted forecasters to the likelihood of severe storms. According to Jim Johnson, chief of the Dodge City, Kansas, forecast office, "the computer models failed miserably with this event [while] data from the profilers . . . tipped the scales toward a successful forecast. What had begun as a fairly low risk for severe storms escalated to an extremely dangerous situation. Fortunately, staff at the Storm Prediction Center . . . used the profiler data early in the analysis cycle and realized the gravity of the situation. If they had trusted model forecasts they would have badly under-forecast a major tornado outbreak with possibly much greater loss of life." Although the Wind Profilers are still an "experimental" program, National Weather Service forecasters consider them indispensable.

+   +   +

Completeness demands at least passing mention of a "nowcasting" tool prized by power company engineers and wildland fire officials as well as weather forecasters. Lightning detection networks can discover flaws in transmission systems, pinpoint likely locations of forest fires, and identify severe thunderstorms fifteen minutes ahead of radar. A direct threat to power lines and parched woodlands, lightning strokes yield dramatic images that can signal a storm's development, reveal its trajectory, and herald its decline. Among forecasters worried about sluggish response by a overwarned public, knowing when a warning is unnecessary is nearly as important as knowing when to broadcast an alarm.

Unlike other monitoring methods examined in this chapter, ground-based lightning detection systems estimate a lightning flash's position by comparing measurements for multiple locations. Cloud-to-ground lightning generates a magnetic wave that propagates outward at the same rate in all directions. One approach, called direction finding, estimates the ground-strike location from the angles at which the magnetic wave arrives at two or more sensors (fig. 5.9, left). Although triangulation principles would have two lines of arrival intersect at the strike point, a band of uncertainty of roughly 1 degree surrounds each azimuth. An alternative method calculates the strike location by comparing times at which the magnetic wave arrives at three or more sensors (fig. 5.9, right). For each pair of sensors, arrival times define a parabola of possible strike locations, and a third station defines two additional parabolas, which in principle should intersect at the exact strike location. Although accuracy depends on network density and the number of sensors contributing to the average estimate, time-of-arrival detection is generally more precise than directional triangulation.

The National Lightning Detection Network, initiated in the early 1980s by a consortium of electrical utilities, uses a mixture of direc-

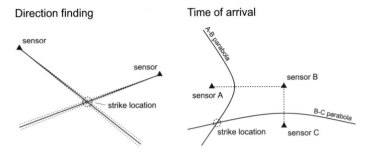

FIGURE 5.9 Lightning detection systems estimate ground-strike location from the estimated azimuths of magnetic waves *(left)* or from time-of-arrival parabolas calculated for pairs of detector sites *(right)*.

tion-finding and time-of-arrival sensors. Linking more than one hundred sensor sites scattered across forty-eight states, the network estimates stroke locations to within roughly 500 meters. Although lightning is sudden—a typical flash consists of four strokes each lasting approximately 30 microseconds—stroke data are usually displayed in one-minute chunks or aggregated into more visually stable images covering periods ranging from five minutes to an hour. To get a sharper picture of a storm's location and intensity a National Weather Service forecaster might overlay fifteen-minute lightning-stroke data on an infrared satellite image showing precipitable moisture or cloud height.

+   +   +

Meteorological surveillance systems complement one another in myriad ways. Doppler radar affords more locally detailed pictures than geostationary satellites, which in turn provide continuous coverage of large storms over land and water. Similarly, Wind Profilers offer a closer, bottom-up view of air flow at different levels and furnish valuable guidance to computer models (to which GOES sensors also contribute), while ground-based lightning detection systems monitor electrical storms and tell wildfire officials where to look for future forest fires—a day later, perhaps, after smoldering tree roots ignite surface litter. What's more, systems for watching weather contribute to other surveillance efforts, such as monitoring world agriculture, pinpointing downed aircraft and foundering ships, and understanding global climate. This complementarity includes GPS satellites: because the GPS time signal is measurably retarded by atmospheric moisture, a ground station that knows the right time can estimate the amount of intervening water vapor.

Given this diverse array of relevant sensors and imagers, it's hardly surprising that the most promising—and for a time the most troublesome—part of the National Weather Service's modernization program is the Advanced Weather Interactive Processing System (AWIPS), an intelligent graphics console designed to help forecasters overlay images, explore data, and compare models. Because diverse views of the atmosphere are the key to accurate prediction, the power of meteorological surveillance depends on the forecaster's ability to integrate information.

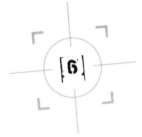

[6]

Wire Loops and Traffic Cams

Traffic signals can be maddening, particularly on lightly traveled roads when the light turns red as you approach. Had
the signal stayed green a second or two longer, there'd be no
need to stop, watch a handful of cars clear the intersection,
and wait that decade-long half-minute for the green light.
Equally frustrating is the certainty that a traffic cop (if we
could afford to replace machines with people) would have already let the waiting cars cross or turn left so that no one,
neither you nor they, need stop and wait. If we can send men
to the moon, why can't we make a smart stoplight?

The answer is that we can. Intelligent traffic signals are
expediting traffic through congested inner cities and allevi-
ating aggravation in the suburbs. In addition to Advanced
Traffic Management Systems (ATMS), which include more

attentive stoplights, Intelligent Vehicle/Highway Systems (IVHS) technologies promise to increase traffic flow by making driving more automatic. Updating that damned traffic signal out on Route 32 is only a question of time and money.

There's more to the story than smart stoplights. Traffic control systems need to know where the cars and trucks are, how swiftly they're moving, and where accidents and breakdowns are disrupting the flow. And because successful modeling depends on predictable behavior, control systems need to know who's undermining the computer model by running red lights or driving aggressively. As motorists in several states are well aware, video cameras can photograph drivers and license plates, electronic character recognition systems can match plate numbers with registration data, and violations tracking systems can send out bills with a picture of the driver caught in the act. What's more, automatic toll collection, a form of electronic surveillance that reduces congestion at toll plazas, might eventually allow authorities to finger dangerous drivers as well as decrease traffic and raise revenue by charging motorists for using city streets during rush hour. An added cost of this automotive utopia is the privacy lost when computers know who we are and where we've been.

+   +   +

At their most rudimentary level, traffic lights have two options: preset timing and traffic-activated interruption. A simple traffic-activated signal would favor the more heavily traveled "major" street with a steady green light except when a motorist approaching along the intersecting "minor" street activates a detector, which changes the signal long enough for a plausible number of waiting vehicles to cross or turn. Used mainly at mid-block pedestrian crossings and entrances to factory parking lots, traffic-activated interruption is the electronic equivalent of a near-sighted traffic cop. More widespread is the clock-driven signal with a programmed green-yellow-red cycle favoring the major street's larger volume with a proportionately greater "split" (longer green light). Although time-of-day adjustments can expedite rush-hour traffic and a more intricate cycle can accommodate left turns and pedestrians, the noninteractive pretimed signal is similar in concept to the light timers used to discourage burglars when we're away.

The effectiveness of a pretimed stoplight depends on predictable traffic and accurate vehicle counts. Before installing a new signal, traffic engineers collect data for several days with portable clock-driven counters housed in kid-proof boxes chained to telephone poles or trees along approaches to the intersection. A flexible pneumatic tube less than an inch in diameter extends outward into the road perpendicular to the curb line. Whenever a wheel rides over the tube, the counter senses increased pressure and records another axle. Anchored to the pavement with large staples, the tube stops short of the yellow dividing line so that the counter registers axles traveling only one way. Although unable to distinguish cars from other vehicles, pneumatic counters are generally more reliable and less expensive than human counters, who become bored or need to pee.

Differential counts provide a rational basis for offering one street a bigger split or adjusting the timing for rush-hour traffic. If turning vehicles require special attention, traffic engineers can place additional counters downstream from the intersection or, in the case of a dedicated left-turn lane, use two upstream counters, one with a shorter tube to capture only vehicles going straight or turning right. The number of vehicles turning left can then be estimated by subtracting the short-tube count from the long-tube count.

Along a heavily traveled rush-hour route, synchronization obviates the need to count vehicles at every intersection. The "time-space diagram" in figure 6.1 describes a typical scheme for coordinating signals so that northbound traffic along a hypothetical Main Street encounters few if any red lights. The vertical axis shows distance, the horizontal axis represents time, and the straight lines sloping upward describe a constant speed of 27.3 miles per hour—the speed at which a vehicle covers the 800 feet between intersections in exactly 20 seconds. The plan shown here calls for a green-yellow-red cycle of 60 seconds. For each signal-controlled intersection a segmented horizontal bar describes the cycle's division into 30 seconds of green (shown in white) followed by 3 seconds of yellow (in gray) and 27 seconds of red (in black). Especially important is the "offset" of the green phase from the green light at 1st Street: synchronized offsets provide a succession of 30-second-wide "through bands" (the white diagonal stripes in fig. 6.1) allowing unimpeded passage for vehicles traveling at 27.3 miles per hour.

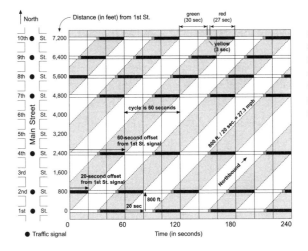

FIGURE 6.1 Time-space diagram for synchronized traffic signals along a hypothetical one-way street with evenly spaced intersections.

In figure 6.1, each white stripe's leading, right-hand edge represents a "green wave" that advances northward as signals change from red to green. Drivers who adjust their speed to the green wave will average 27.3 miles an hour and not need to stop or slow down. When everyone cooperates, synchronization can move many vehicles along an otherwise congested arterial. If Main Street is one-way and four lanes wide, for example, a hypothetical "platoon" of eighty cars, 15 feet long and 45 feet apart, could cover the 1.4 miles from 1st Street to 10th Street in three minutes without hitting a single red light. And even if breakdowns, buses, mid-block stops, and slowpokes disrupt the dance, trailing a green wave beats halting at every second or third traffic light.

Straightforward synchronization makes it easy to understand why traffic engineers love one-way streets. If only three lanes are available, for example, dedicated left- and right-turn lanes at intersections with two-way streets allow at least one dedicated lane for through traffic, which is more predictably uniform. And if the cross streets are also one-way, a second through lane encourages even greater predictability. Moreover, the coordination plan in figure 6.1 adapts readily to unevenly spaced cross streets—note that by shifting the green segments to the right or left, any of the horizontal bars may be moved up or down without affecting the 27.3 miles-per-hour speed or the 30-second bandwidth. What's more, an engineer can easily adjust the speed of the green wave as well as the length of the cycle and its constituent phases by changing the slope and width of the through bands.

Two-way traffic is more complicated. Figure 6.2 illustrates the challenge of coordinating the signals in figure 6.1 for two-way traffic. In this example, twin sets of through bands intersect at cross streets and treat both directions equally. My solution retains the 27.3 miles-per-hour speed but reduces the cycle from 60 to 40 seconds and the green phase from 30 to 20 seconds. A yellow phase of 3 seconds leaves only 37 seconds for green and red combined. Adding an all-red phase—an essential safeguard against aggressive drivers who race into the intersection when the light turns red—would further constrain the trade-off between green and red. The 40-second cycle, which reflects the 800-foot distance between intersections, might be overcome by placing signals only at every second or third intersection—or by allowing an occasional red light along one or both directions. Where cross streets are not evenly spaced, two-way synchronization is even more problematic. For unimpeded green waves in both directions the simplest, most workable strategy is one-way traffic along Main Street with a parallel street handling the other direction.

Irregularly spaced intersections are no less troublesome than the repeated retiming studies needed to synchronize cycles, phase lengths, and offsets as well as accommodate left-turn lanes, pedestrian crossings, transit vehicles, railroad crossings, and emergency vehicles. Although ongoing traffic counts can help transportation officials deal with commuter traffic, seasonal effects, special events, and bad weather, a more robust strategy is to collect flow data in real time and let a network of remote detectors, communication lines, and centralized computers adjust signals to actual conditions. Dynamic adjustment is especially important where major streets intersect.

Automated traffic management can also ease congestion on freeways by coordinating entrance-ramp signals, which regulate access, and variable-message signs, which advise motorists to change lanes, slow down, expect delays, or take an alternate route. Because it's prohibitively expensive if not politically impossible to add another expressway, Los Angeles and other sprawling metropolitan ar-

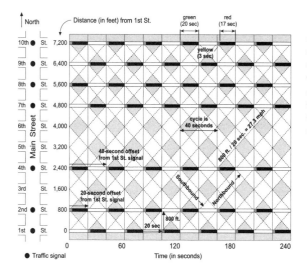

FIGURE 6.2 Time-space diagram for synchronized signals with equal accommodation of two-way traffic along the hypothetical street in figure 6.1.

eas with dim hopes for public transit look to ATMS for solutions to air pollution and other environmental consequences of rampant postwar road building.

+   +   +

Whether connected to a central computer at the traffic control center or to a microprocessor in a nearby signal cabinet, traffic-activated signals depend on electronic or electromechanical "detectors" designed to count cars, estimate speed, or merely sense the presence of vehicles waiting to cross or turn. Although the detector might be a pressure-sensitive pad, a microphone positioned to pick up engine noise, or a small radar system suspended over the roadway on a pole, most municipal traffic engineering departments rely heavily if not entirely on "inductive loop detectors" buried in the pavement. The loop is a few turns of a thin, insulated electric wire, which carries an alternating current that changes direction between ten and two hundred times a second thereby setting up a magnetic field that oscillates at the same frequency. Through a phenomenon called the ferromagnetic effect, this magnetic field induces a similar but opposite magnetic field in any nearby metallic conductor. Because the second magnetic field absorbs magnetic energy from the loop, the detector senses a vehicle whenever a large metallic object, such as an engine block, disrupts the current in the loop. A technician calibrates the loop so that when a disruption exceeds the threshold, the detector reports the presence or passage of another vehicle.

Loop detectors are easy to recognize. To embed the wire, installers cut a series of straight slots in the asphalt or concrete with a rotary saw. To avoid right-angle bends, which can stress the wire, they often bevel the corners with short, 45-degree cross cuts, as in figure 6.3. A longer cut accommodates a lead-in wire, which links the loop to a "pull box" buried in the ground at the side of the road and connected to the signal controller. Cuts are no more than a half-inch wide and about 1 to 3 inches deep. After placing the wire in the slot, the field crew fills the cut with an epoxy sealant, which encases the wire, keeps out moisture, and restores the road surface. Passing tires burnish the epoxy to a glossy gray finish, which sometimes suggests metal strips driven sideways into the pavement.

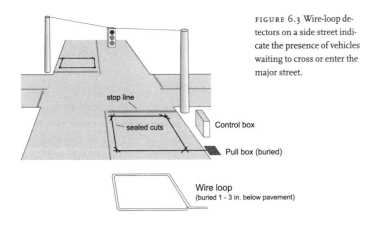

FIGURE 6.3 Wire-loop detectors on a side street indicate the presence of vehicles waiting to cross or enter the major street.

stop line

sealed cuts

Control box

Pull box (buried)

Wire loop
(buried 1 - 3 in. below pavement)

If you're curious about a signaling system's priorities, look at the placement of its loop detectors. A convenient reference point is the stop line, usually a wide white stripe perpendicular to the curb line and stretching halfway across the road just upstream from the pedestrian crossing. A loop along a minor street immediately in front of the stop line is probably intended to detect the presence of a waiting vehicle—if nothing's waiting, heavy traffic along the major street can enjoy a longer green. By contrast, a rectangular loop extending downstream from the stop line well into the intersection can sense vehicles waiting to cross or turn left—situations in which a longer yellow or left-turn (green-arrow) phase can avert gridlock or the frustration of waiting several cycles to make a left turn. At key intersections separate loops usually monitor the left-turn and through lanes.

Detector loops are especially useful along freeway ramps. Loops at both ends of an on-ramp can estimate the number of waiting vehicles, if any, whereas a detector midway down a long off-ramp can activate the stoplight ahead, which might fortuitously turn green just as your car reaches the end of the ramp. And a detector farther upstream might warn of a dangerous queue, which could back up onto the freeway unless the controller clears cars off the ramp more rapidly.

Loop detectors are usually designed for specific tasks such as counting moving vehicles or noting a stopped car or truck. The shape and size of a loop affect the extent and sensitivity of its magnetic field. For example, diamonds are deemed less likely than other designs to pick up vehicles in adjacent lanes—traffic engineers call this "splashover"—while long, narrow rectangles centered on each lane's oil line are more suitable for moving traffic. Reliable counts are especially important when transportation departments evaluate the durability of pavement or assess the need for new or wider roads. Bicycles and small motorcycles, with less inductive presence than a car or truck, are especially dicey. To detect small vehicles, loops at traffic-activated stoplights sometimes include a smaller subloop, with a few extra windings of the wire just shy of the stop line. Toronto and other cyclist-friendly cities use a series of three or four white dots to mark small bike-sensitive loops so that savvy cyclists will know where to stand when they want to cross.

A pair of closely spaced loops in the same lane can measure speed as well as traffic volume. Although traffic cops rarely use loop detectors to catch speeders—radar guns are less expensive and more portable—fear of speed traps was especially troublesome in the 1960s, when embedded loops were relatively new and some anxious truck drivers would hit the breaks at the sight of telltale lines of epoxy sealant. The worst enemies of a wire embedded in asphalt are heavy vehicles that stop suddenly and wrinkle the pavement.

Because sudden stops also cause rear-end collisions, traffic engineers are especially concerned about the signal that turns yellow when a driver is too close to the intersection for a smooth, safe stop. A motorist who cannot stop easily but is too far away to clear the intersection before the light turns red is in a "dilemma zone" where indecision and panic could precipitate a crash. An all-red phase can give the driver who doesn't stop added time to clear the intersection, but the occasionally fatal consequences of a sudden stop call for delaying the yellow phase when a vehicle is still in the dilemma zone. Although an extended green is not always practicable for steady traffic—drivers queued up to turn or cross cannot wait indefinitely—properly placed loop detectors can prudently delay the yellow phase if no one has been waiting long. Some signal controllers get by with a pair of detectors, one on each side of the zone, but more advanced systems require multiple detectors to monitor closely the number and speed of approaching vehicles. Speed is important because a fast vehicle has a long dilemma zone whereas a slow one might not need an extension.

However widespread, wire loops are not the final word in smart signaling. Some cities are supplementing or replacing their loop detectors with video cameras connected to machine-vision processors that count cars, detect queues or stopped vehicles, and differentiate inbound from outbound traffic and through vehicles from those making turns. A single camera can cover several lanes, and when positioned strategically along highways, video detectors can also monitor speed as well as spot accidents, breakdowns, and other "incidents." More expensive initially, video systems are easier to install and maintain, especially where heavy trucks tear up the pavement. In the Texas Panhandle, for instance, loop detectors often last less than eighteen months and cost as much as $15,000 to replace. As Texas Department of Transportation official David Miller told me, "Because wires embedded in asphalt can fail at any time, no one guarantees loop detectors." By contrast, contractors charge about $20,000 to install four cameras and their electronic controller, and the cameras come with a two-year warranty.

Machine vision is largely a matter of rapid, conscientious number crunching that simulates human vision by detecting edges, matching patterns, and estimating displacement. Figure 6.4 de-

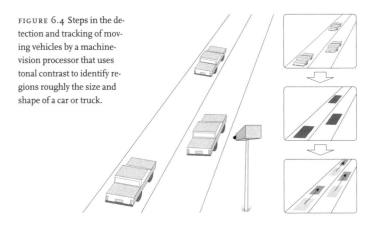

FIGURE 6.4 Steps in the detection and tracking of moving vehicles by a machine-vision processor that uses tonal contrast to identify regions roughly the size and shape of a car or truck.

scribes the steps. A computer chip in the traffic controller scans a scene, notes tonal contrast, and outlines regions that differ from the pavement and are roughly the size and shape of a car or truck. Once calibrated, the system can easily relate an object's position in the image to its location on the ground. And by tracking the object through successive scenes captured a fraction of a second apart, the processor can measure a vehicle's speed as well as detect cars making turns or changing lanes. By covering a broad area, a single camera avoids the splashover that can occur when a car occupies parts of two lanes. And if an accident occurs or a queue develops, the processor can report the incident over a telephone line or radio channel. Except during installation, the machine-vision processor does all the watching.

+   +   +

Not all traffic video cameras feed machine-vision processors. In Syracuse, New York, where I live, fiber-optic cable connects the city's traffic control center to cameras at five intersections. Officials selected these locations because of heavy traffic or a history of accidents. The control room has a huge TV screen—the kind found in sports bars and corporate boardrooms—and three smaller monitors. On the day I visited, one of the smaller monitors was cycling through the five cameras. Control room operator Harry Carlson showed me how he can call up any camera on any monitor, pan around with a joystick, and zoom in for greater detail. By entering a number, he can also rotate each camera to a carefully selected, pre-programmed "shot" showing traffic approaching or leaving the intersection in a specific direction. Mounted 30 feet above the ground, the cameras can focus on incidents more than a block away.

Syracuse's closed-circuit TV cameras are part of a larger system that lets Carlson monitor and, if necessary, reprogram traffic signals at 143 of the city's 170 signal-controlled intersections. Control boxes at each intersection report in every 30 seconds, and a server stores their data for analysis by MIST (Management Information System for Traffic), a software package that optimizes offsets and extensions for through traffic in multiple directions. For hands-on analysis, a graphics workstation displays an interactive map of the area under central control. Numbers next to several dozen street segments represent volume, occupancy, or average speed, and a pull-down menu lets the operator change indicators in an instant. By clicking on an intersection, Carlson can retrieve a detailed dynamic diagram showing signal conditions in real time for each driving lane and pedestrian crosswalk. The intersection diagram also shows the presence or absence of vehicles at each loop detector and indicates whether a pedestrian has activated a "call switch." Because intersection diagrams occupy only a fraction of the screen and can be dragged about, the operator can call up several intersections at once and reposition their diagrams in the correct sequence along a synchronized route. When the maps indicate unusually sluggish traffic, the closed-circuit TV cameras might suggest a likely cause and an appropriate response.

The Syracuse Signal Interconnect Project cost over $11 million, largely to provide underground communication lines. "It's very expensive to open up the streets and bury several miles of fiber-optic cable," Carlson explained. Indeed, the cost of cabling is the primary reason why intersections well away from downtown are not linked to the central computer. Although centralized synchronization has noticeably reduced average waiting time at stoplights—by as much as 50 percent, according to one study—$11 million is a lot of money to save commuters a couple of minutes of frustration each day, especially for a medium-size city like Syracuse. The real motivation was air quality: vehicles idling at intersections threaten lungs and hearts by spewing out tons of carbon monoxide and other pollutants. That's why the federal government picked up 80 percent of the tab, and New York State subsidized another 15 percent. Synchronized stoplights are about more than saving time.

+ + +

Bigger cities have another job for traffic surveillance: helping commuters select a route or departure time that avoids congestion. From metropolitan Washington, D.C., to Seattle, Washington, transportation departments have set up Web sites that offer continually updated maps of traffic conditions as well as live views from strategically placed video cameras. Although the effect on driver behavior is difficult to gauge, the added cost is modest because control room personnel need the data anyway to monitor flow, regulate signals, and deal with emergencies. Sharing the information helps motorists assess their options and lets the department tout its efficiency and commitment.

In Washington State's Puget Sound area, for instance, the Department of Transportation Web site (www.wsdot.wa.gov/PugetSoundTraffic/) lets commuters choose a north-up or west-up overview. Additional maps provide greater detail for the city of Tacoma, the bridges across Lake Washington (fig. 6.5), and the northern and southern approaches to Seattle. Focused on expressways, the maps describe traffic conditions in each direction for segments roughly a mile long on the overview maps and a half-mile long on the detailed maps. Mimicking a stoplight, red, yellow, and green symbols differentiate "heavy," "moderate," and "wide open" traffic, respectively, while ominous black segments denote "stop and go" conditions, which commuters want to avoid. Gray and blue identify the network's remaining segments as "no data" (detector out of service) or "no equipment" (detectors not installed). A separate key at the top of the map shows current traffic directions along the network's reversible "express lanes"—parallel roadways that increase capacity for portions of the I-5 and I-90 corridors by changing direction with the tide of commuter traffic. Separate symbols describe conditions along the express lanes, which have no direct interchanges with cross streets and are less likely to experience back-ups near off-ramps.

Motorists curious about distinctions between "heavy" and "stop

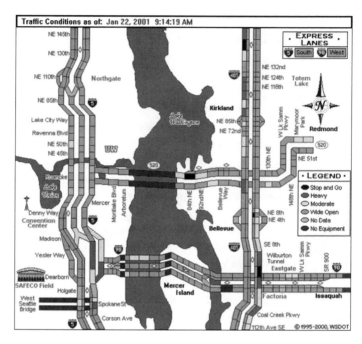

Traffic Conditions as of: Jan 22, 2001 9:14:19 AM

FIGURE 6.5 Detailed "Bridges" map focuses on freeways around Seattle's Lake Washington. This late-morning view shows isolated stretches of "heavy" and "stop and go" traffic. The directional key at the upper right shows the express lanes for I-5 and I-90 aligned for traffic moving to the south and west. From Washington State Department of Transportation, "Puget Sound Traffic Conditions," http://www.wsdot.wa.gov/PugetSoundTraffic/.

and go" can click on the Web site's "Cameras" button, which pulls up an index map showing locations of eighty "traffic cams" in the Seattle area. Each icon is a clickable symbol, or button, that summons a small, low-resolution video snapshot of current traffic (fig. 6.6). Viewers can quickly infer driving conditions from the average spacing of vehicles and the number of open lanes—if cars are bumper-to-bumper, that's clearly bad news. The images reload automatically every 90 seconds and are transmitted over the telephone network, which is less expensive than dedicated fiber-optic cable. Additional index maps show locations of seven traffic cams in the Tacoma area, five cameras focused on critical street intersections in Seattle, and twelve cameras covering local traffic in Bellevue, a municipality east of Lake Washington that touts itself as "the nation's most wired city."

Washington transportation officials put their traffic cams online in 1996, in response to requests from area residents who had seen the images on local television. Less than half of the state's two hundred cameras feed the Web site, which cost only $20,000 plus a bit of employee time to set up. Web traffic increased markedly in December 1999, when demonstrators at the World Trade Organization meeting disrupted Seattle traffic. Between November 29 and December 6, for instance, the number of visitors jumped from 8,000 to 18,000 per day. Severe weather also stimulates usage. Residents without Web access can monitor traffic flow on television. Tacoma's city-run information channel added the local traffic cams to its morning and evening rush hour programming in 2000, and two other cable services carry the images. Don't expect to check out serious accidents, though: because live television might disturb victims' families, TV Tacoma switches quickly to its events calendar.

Traffic cams and automated surveillance offer intriguing opportunities for collaboration among public agencies and private firms.

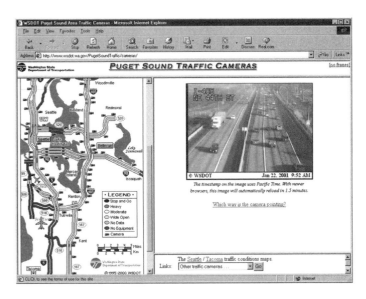

FIGURE 6.6 Clickable camera icons on the index map *(left)* summon traffic cam images *(right)*, which update every 1.5 minutes. From Washington State Department of Transportation, "Puget Sound Traffic Conditions," http://www.wsdot.wa.gov/PugetSoundTraffic/.

In metropolitan Washington, D.C., for example, SmartRoute Systems, a Cambridge, Massachusetts, firm, formed a partnership that includes the *Washington Post*, the local NBC affiliate, and transportation departments in Maryland and Virginia. Named Partners in Motion, the consortium offers free SmarTraveler bulletins on the newspaper's and television station's Web sites. Reports include verbal alerts about delays and hazardous conditions, current driving times between key points, and a clickable map showing camera locations along the Capital Beltway. Unlike Puget Sound's traffic snapshots, some of the D.C. traffic cams refresh the image several times a second. Additional buttons point to Web sites maintained by transportation agencies in Maryland and Virginia, which maintain over two hundred cameras throughout the area and supply basic data to the local SmarTraveler control center. In addition to the Web sites subsidized by advertisers or taxpayers, SmarTraveler and other traffic reporting companies are exploring a variety of fee-based services, including customized traffic reports based on the subscriber's normal commuting route and delivered automatically to a pager, cell phone, onboard navigator, or wireless handheld computer. Although timely traffic data are highly useful, wireless delivery of tiny maps to distracted drivers might have unintended consequences.

+ + +

Photo surveillance is taking on another of the traffic cop's responsibilities—ticketing motorists who run red lights. A camera mounted above the intersection takes a snapshot of any vehicle entering after the signal turns red. These cameras typically use photographic film, which must be collected manually but provides a clearer picture than video. A clerk or image processor then reads the license plate so that a computer can find the owner's name and address in the registration database and print out a citation, complete with date, time, location, and direction of travel. The more sophisticated stoplight-camera systems photograph the windshields of approaching vehicles and embellish the ticket with a picture of the driver. Graphic evidence makes it hard for the guilty motorist to claim it's all a mistake—and easy for a parent to ground the son or daughter who ignores the rules. Video surveillance also helps localities crack down on motorists who speed through school zones or dodge around gates at railroad crossings. That said, localities willing to reduce violations (and revenue) could do so by extending the yellow phase a second or two.

Because a machine, not a police officer, generates the ticket, photo enforcement typically requires special legislation and a different set of rules. Most states do not levy points, which can increase the violator's insurance rate, and some call the mailed-out citation a "notice of liability." Although the practice might seem heavy-handed, especially when $100 fines help pay for $20,000 cameras, traffic safety experts point to an alarming increase in the number of American drivers who run red lights and cause 260,000 traffic accidents every year, about 750 of them fatal. Studies indicate that four-fifths of the population supports photo enforcement—nearly 80 percent according to a recent report by the Insurance Institute for Highway Safety—and red-light running declines markedly when motorists know they're likely to get a ticket. And because perceived risk is at least equally important, Toronto, Canada, ran an extensive advertising campaign and posted large signs with a camera logo and the warning "Red Light Camera in Operation" (fig. 6.7). Most motorists, I suspect, are unaware that the city cut costs by buying only ten cameras to rotate among forty intersections.

FIGURE 6.7 Toronto mayor Mel Lastman, a strong supporter of photo enforcement, posed next to this warning sign for a September 1998 press release. After a pilot study at this intersection revealed an average of sixty-five offenses every twenty-four hours, the Ontario government approved the use of red-light cameras. From City of Toronto, http://www.city. toronto.on.ca/mayor/.

Photo enforcement does not sit well with libertarians, privacy advocates, and political conservatives who see the cameras as yet another encroachment of Big Brother. Inherently suspicious of video surveillance and databases, the American Civil Liberties Union is especially wary of technologies that integrate cameras and computers. As Michael Klein, an ACLU board member told the *Ventura County Star,* "What's going on right now is an increasing ability of the government to control your life. And cameras are a part of it." Resistance can be strident and effective. California legislators rejected photo enforcement after a bitter, highly partisan battle led by conservative Republicans like Assemblyman Bernie Richter, who sees red-light cameras as a form of "general surveillance [that] smacks of Nazi Germany in the 1930s." Yet many liberals and law-and-order conservatives think the end—fewer crashes, fewer deaths—justifies the means. And as Howard County, Maryland police lieutenant Glenn Hansen points out, "It's not a new law; it's a new way of enforcing the law." My hunch is that public consensus lies in selective enforcement—ignoring routine speed violations (which the police largely overlook anyway) and adjusting the size of fines to focus on persistent offenders—and camera systems that photograph only violators. Big Brother is less a threat to innocent drivers if he lacks long-term memory.

+ + +

The story of traffic surveillance doesn't end here. Geographic technology can readily monitor our comings and goings in far greater detail than loop detectors and stoplight cameras. You know what I mean if you use E-ZPass, a clever electronic system for paying road and bridge tolls in several northeastern states. A vehicle approaching a toll barrier slows to five miles per hour and passes a detector that directs the Automatic Vehicle Identification (AVI) system to query the tiny transponder, or "tag," attached inside the windshield at the top. Sealed in a watertight case with a lithium battery that lasts ten years, the tag is a tiny transceiver, which sends back its unique serial number so that a central computer can debit the owner's account. For road tolls, which are based on mileage, the system compares entry and exit points and reports the details on your monthly statement together with the date and time of exit, down to the second. If the goal were speed enforcement, E-ZPass could easily calculate your elapsed time and—if you drive like most of us—send out a "notice of liability."

Electronic toll collection is the precursor of a transport revolution that is at once hopeful and ominous. Market-based road pricing schemes, popular among avant-garde urban planners in Europe, promise to alleviate traffic congestion and air pollution by charging drivers for using city streets during rush hour. Bolstered by GPS and E-ZPass-like transponders, AVI technology could automatically enforce a host of traffic regulations, from speeding and red-light running to double parking, tailgating, and incomplete stopping at stop signs. A markedly different control strategy focuses on roads and cars rather than drivers. "Smart cars" with sensors that keep them safely apart and in their proper lanes could cut down on collisions as well as increase throughput along existing corridors. Where dispersed metropolitan populations and highly varied patterns of origins and destinations thwart efficient public transport, drivers using local roads could converge on grouping stations at which vehicles would be assembled into closely spaced, train-like packs for high-speed travel to downstream stations and branch points. "Drivers" would sacrifice direct control over their vehicles for much of the trip—along with the hazards of road rage and multi-car pile-ups—and could read, watch television, or talk safely on cell phones.

Although skeptics question the economic benefits of automated highways, some heavily automated, highly regimented scenario seems inevitable, at least in the long, long run. And even if exponential population growth and sprawling metropolises ultimately force urban commuters off the road onto public transit, routine surveillance of passengers and vehicles will be a key component of whatever systems transport us into the next century. Equally certain is the ability of a highly automated transportation infrastructure to monitor movement and control lives in ways few American alive today would find acceptable. Although this expanded capacity for surveillance does not make intrusive regimentation inevitable, our Brave New World's Brave New Roadways will surely challenge conventional notions of personal privacy.

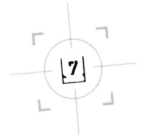

Crime Watch

Essays on privacy and law enforcement almost always bring up the Panopticon, an intriguing blend of surveillance and prison architecture designed by Jeremy Bentham (1748–1832). Bentham was a legal reformer, polemicist, and early advocate of utilitarianism, a philosophy that cedes the moral high ground to whatever works for the greatest number of people. In 1791, he published what remains his most famous work, *"Panopticon": or, the Inspection-House; containing the idea of a new principle of construction applicable to any sort of establishment, in which persons of any description are to be kept under inspection; and in particular to Penitentiary-houses, Prisons, Houses of industry, Workhouses, Poor Houses, Manufactures, Madhouses, Lazarettos, Hospitals, and Schools; with a plan of management adopted* [sic] *to the principle.* This tortuous title is a sanitized description of the design's intent: controlling inmates of various kinds through a constant threat of random surveillance. A drawing (fig. 7.1) described the relationship between the inmate's cells, which ringed a central tower and an "inspection gallery" from which an unseen "inspector" could observe the cell's occupants at any time through visual baffles like venetian blinds. This similarity to closed-circuit television (CCTV) makes Bentham's plan an attractive metaphor for social philosophers who write expressively about the "panoptic gaze" of technologies of social control or the "panoptic power" of electronic surveillance.

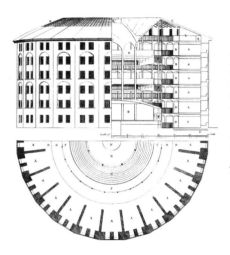

FIGURE 7.1 This drawing of Bentham's Panopticon shows the annular arrangement of cells around the inspection galleries, which are connected through a central tower to the "Inspector's Lodge" below. From Jeremy Bentham, *The Works of Jeremy Bentham, published under the superintendence of his executor, John Bowring* (Edinburgh: William Tait, 1843), vol. 6, following p. 172.

Although the metaphor seems vaguely appropriate, I am not convinced that the similarity between Bentham's model prison and video surveillance tells us anything that's not obvious about the watcher's power over the watched. My hunch is that the prison's walls and bars as well as the isolation of inmates in individual cells exert far greater control over prisoners' lives than a ready ability to spy on their actions. No doubt a voyeur-inspector would enjoy a stronger sense of power and control as well as an increased ability to make the inmates docile by rendering swift if not severe punishment for minor infractions—at least occasionally and with sufficient retribution to indicate which rules are not to be broken. What's relevant, though, is the power of surveillance to intimidate someone *already* under the watcher's control, like a prisoner (who can be beaten), an employee (who can be fired), or a motorist who runs red lights (and could be fined or lose his or her license). What the prison metaphor fails to capture is the ability of video surveillance to monitor behavior at a distance over a wide area and if response is rapid, to protect people from human predators. Remove the *con*, though, and the resulting adjective *panoptic*—*pan* means "all" and *optic* refers to seeing—seems especially appropriate when coverage is broad and thorough.

+   +   +

Video cameras are abundant if not ubiquitous. Office buildings and shopping malls have them to help security staff watch for suspicious behavior throughout the parking garage, stairwells, and obscure parts of the lobby, and banks believe that a visual record of teller and ATM transactions cuts down on armed robbery and debit card fraud. Self-service gasoline stations rely on CCTV to prosecute people who fill up and drive off without paying, while convenience store operators believe that video surveillance can deter impulsive acts like sneaking out with a jar of Cheez Whiz under your sweater. The monitors are often in plain sight: if we know we're being watched, we'll behave ourselves.

In the 1990s video surveillance came out of stores and into our streets and parks—what geographers call public space. Because public officials are skittish about being labeled Big Brother, public-space video surveillance is generally less obtrusive than the cameras used to catch motorists who run red lights. And it's used less widely in the United States than in Great Britain, which began experimenting with CCTV to combat crime and terrorism in the mid-1980s, after IRA frustrations boiled over from Northern Ireland into Central London. In the aftermath of the September 11, 2001, attacks on the World Trade Center and the Pentagon, America might decide to catch up.

For the few U.S. cities with video surveillance of pedestrian areas, coverage seems spotty, typically with less than one camera per block. A survey by the California Research Bureau, a division of the California State Library, reported a variety of video surveillance strategies in use by thirteen municipalities ranging in size from Baltimore, Maryland, and San Diego, California, to Dover, New Jersey. In general, cameras that can pan, tilt, and zoom are connected by fiber-optic cable to police headquarters or another central location, where police officers watch the monitors for part of the day. But some cities rely heavily on public funds or use volunteers, and a few systems merely record scenes passively on tape for periodic review and possible use in court.

Baltimore, for instance, installed sixteen fixed-position cameras in June 1996 in a downtown retail area with a reputation tarnished by auto break-ins, drug dealing, prostitution, and aggressive panhandling. Funded by a federal grant as well as private funds, the cameras are actively monitored by police—in two shifts, 7 A.M. to 11 P.M.—from a nearby 8-by-12-foot kiosk. The system stores its images on videotape, which is changed every twenty-four hours and retained for up to four days, after which tapes are reused or discarded unless needed as evidence in a police investigation or criminal prosecution. Because persons recorded are in plain view and have no expectation of privacy, capturing their movements on tape does not breach the Fourth Amendment ban on "unreasonable search and seizure." And as an added safeguard, only law enforcement personnel watch the monitors and screen the tapes.

Despite such safeguards, privacy advocates object vigorously to systematic monitoring by the police. In October 1996, for instance, after New York City officials vowed to jump aboard the CCTV bandwagon with twenty-four-hour video surveillance in Central Park, Norman Siegel, executive director of the New York Civil Liberties Union, warned that "if we start going in this direction, the next logical extension is to put cameras on every street corner and allow the government to monitor people engaged in innocent, lawful and confidential activities." Widespread cameras are no less a threat than the resulting tapes, according to Siegel, who charged that the video surveillance program was not only "invasive" but "raises the Orwellian specter of Big Brother government spying on its residents, and compiling a video record of the free movement of individuals as they make their way about the city." Debate quickly focused on the tapes' retention period, with Siegel arguing for erasure within seventy-two hours and the police arguing to keep the images for at least a full week because some crimes might not be reported for several days.

Intent on documenting the extent of video surveillance, NYCLU volunteers mapped the locations of all readily visible video cameras in Manhattan. Five months of fieldwork revealed 2,397 cameras, but an accompanying explanation warned that the map is "far from exhaustive" and omits "many more" cameras "tucked surreptitiously out of the line of vision or small enough to escape detection." Surprisingly, the vast majority of the cameras—2,117, or 88 percent—were privately owned, and 55 of the remaining 270 public surveillance cameras were traffic cameras operated by the transportation department, not the police. Big Brother, if he exists at all, is largely a private cop with tunnel vision and a stiff neck.

Published on the Internet to "raise awareness of the prevalence of video surveillance," the NYCLU map reveals highly uneven coverage. The excerpt in figure 7.2, for instance, shows concentrations of video cameras in front of police headquarters (toward the lower right) and around the Federal Reserve Bank (near the left edge). Elsewhere in the area camera density is comparatively low. For instance, only one camera is evident along Church Street next to the site of World Trade Center (large light-shaded block toward the upper left), which terrorists had attacked in 1993 by exploding a fertilizer bomb in the parking garage. Other parts of Manhattan have less than one camera for every four blocks. What's more, although the number of cameras probably has increased since 1998, when the map was made, video surveillance has less salience now that the United States seems unlikely to mimic Britain's eager adoption of police-operated street cameras. NYCLU leaders still follow technical developments like automatic facial recognition, but their Web site fo-

FIGURE 7.2 Excerpt from Mediaeater, "NYC Surveillance Cameras Project," http://mediaeater.com/cameras. Like many views of Manhattan, the map is oriented with north at the upper right.

cuses on more compelling issues like police brutality and reproductive rights.

Civil libertarians are not alone in considering CCTV a threat to personal privacy. In fall 2000, when my graduate seminar focused on surveillance technology, two of my seven students were highly apprehensive about denser, more intrusive coverage with sharper imagery. Fueling their fears was the British experience, which generated a wave of disapproval in the social science literature, as well as cheaper, smaller, more versatile cameras. While it's tempting to dismiss this anxiety as unfounded technological determinism, I have little doubt that our public officials could find a more dramatic and politically expedient response to a sudden increase in street crime, hooliganism, or terrorism. And venal vendors, heretofore content with the expanding private security market, would hardly object.

Whether CCTV monitoring would ever reach Orwellian proportions in either the United States or the United Kingdom is another matter. Perhaps the reason why American police have not followed their British counterparts down the panoptic path (or slippery slope, if you like) is that video surveillance of public space has proved a feeble defense against determined criminals or terrorists and only a localized inconvenience for other lawbreakers. In short, whether it works depends very much on the meaning of *works*. Evaluation studies are methodologically troublesome because cameras that reduce gang activity, drug dealing, and public urination in their immediate vicinity typically displace some of these activities to other locations, often just out of the camera's range. Proponents have oversold its effectiveness to much the same extent that opponents have overestimated its invasiveness. Whatever the technology's limitations, a majority of Britons accept CCTV as a necessary and appropriate law enforcement tool.

Given time and a huge chunk of public money, surveillance technology could, I am certain, become far more intense, powerful, and invasive. One need not be a science fiction fan to envision a future in which cameras as dense as streetlights feed images to central computers with face-recognition algorithms and biometrics software that match pedestrians to their stored profiles and track their movement through streets and parks. Whether this Orwellian enterprise would be worthwhile is another matter: accurate retina scanning requires far greater resolution than conventional street cameras, and several of the location-tracking technologies examined in the next chapter can protect citizens more reliably and less expensively. Biometrics might prove useful for screening airline passengers at the check-in desk, but the hazards of misidentification—fingering innocent pedestrians while ignoring known terrorists—seem far more daunting than the threat to personal privacy.

+ + +

As the courts see it, whether random surveillance is constitutional usually hinges on whether citizens have a reasonable expectation of privacy. The landmark case is *Katz v. United States,* in which the Supreme Court ruled that FBI agents were wrong in putting a listening device inside a phone booth. A court order can authorize the police to tap your telephone or bug your living room, but detectives can't randomly prowl public space with a highly sensitive directional microphone in hope of uncovering a drug buy or stock swindle. Expectation of privacy is hardly an issue, though, when the sound in question is a gunshot. Residents blocks away hear the noise, and anyone alarmed by armed revelry and not afraid of the culprits might dial 911. Shooting into the air might seem a harmless way to celebrate the New Year and other festive occasions, but the laws of physics dictate the bullet's return to earth with deadly force. If you think people who complain are spoilsports, tell it to the two daughters of thirty-one-year-old Benjamin Velasco. Shortly after 1 A.M. on January 1, 2001, Velasco was headed for a party with his wife when a stray bullet struck and killed him. That morning falling bullets hit five other Los Angeles-area residents, none fatally. Velasco was not the nation's only fatality. At midnight in El Cenizo, Texas, for instance, a falling bullet struck a fourteen-year-old girl in her upper chest while she was standing with her mother in their front yard.

Instead of waiting for a complaint, police in several California cities rely on ShotSpotter, which its inventors describe as an "automatic real-time gunshot locator and display system." Their patent application, filed in August 1997 and approved in October 1999, portrays a clever marriage of seismic analysis and acoustic filtering. Landlines or wireless transmitters connect a network of pole- and

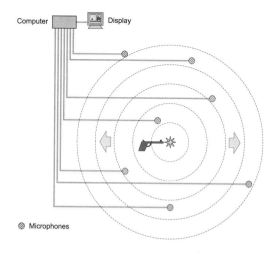

Computer　Display

Microphones

FIGURE 7.3 ShotSpotter calculates a gunshot's location
from the relative arrival
times of the muzzle burst at
a network of microphones
connected to a central computer. Adapted from figure 1
in Robert L. Showen and
Jason W. Dunham, applica-
tion for U.S. Patent no.
5,973,998, awarded October
26, 1999.

roof-mounted microphones, roughly 1,000 to 3,000 feet apart, to a
central computer. Like an earthquake, a gunshot generates a sharply
defined circular pulse, which expands outward at a constant speed.
As figure 7.3 shows, ShotSpotter's microphones detect the wave at
slightly different times depending on their distance from the
shooter's location. The computer can use any three arrival times to
calculate coordinates for a shot's origin and can use intersecting circles or spheres to triangulate a location in either two or three dimensions—especially useful in hilly terrain or for sensors at radically
different heights. And when four or more microphones detect the
same gunshot, ShotSpotter can choose the most reliable triangulation. The process pinpoints gunshots to within 15 yards, which can
narrow the location to a particular house and perhaps one or two of
its neighbors.

A key component is a sophisticated acoustic filter able to distinguish the abrupt muzzle blast from the weaker and less geomet-
rically reliable sound of a bullet. ShotSpotter also differentiates gun-
shot-like explosions from background noise, including the weaker,
more localized sounds of hammering, barking dogs, and slammed
car doors as well as the loud, more continuous sounds of airplanes
and train whistles. In determining arrival times, the system correlates acoustic waves from its various microphones, and it can use
the onset and separations between multiple gunshots to calculate

direction and speed for moving sources typical of "drive-by" shootings. Although able to discriminate muzzle blasts from their reverberations, ShotSpotter also triangulates fireworks, gas explosions, and backfiring vehicles. In alerting the police dispatcher to a possible gunshot, the system displays the location on a map and stores a short sound "snippet" for human confirmation.

Police use the maps as propaganda—graphic warnings telling shooters to cut it out. As the vendor's Web site asks rhetorically, "What would you do if an officer knocked at your door [with] a computer generated map that showed gunfire in your backyard? Chances are you would not do it again! If you did, you know there is a high probability that an officer will knock at your door and potentially someone will be arrested for committing this crime." However intimidating, preventive warnings are less expensive than prosecution and incarceration as well as less traumatic than the physical and emotional injuries of neighbors hit by stray or falling bullets.

Redwood City, California, a suburb of San Francisco, adopted this approach. Concerned about citizens celebrating birthdays and assorted holidays with random gunfire, city officials allocated $25,000 for a ten-week trial in 1995. Eight microphones monitored a 1-square-mile area notorious for noise and falling bullets. Thanks to "Operation Silent Night," the next New Year's was remarkably quiet, and city officials credited ShotSpotter with cutting the monthly average number of gunshot incidents from twenty-four to twelve. Although critics and the police union objected to the cost, Redwood City eventually bought the system outright—after the company reduced the price from $250,000 to $85,000.

Redwood City's success no doubt inspired the Los Angeles County Sheriff's Department to arrange a test in the densely populated Willowbrook area, the scene of gang activity and drive-by shootings as well as random gunshots. In addition to promoting prompt response by street cops, the L.A. test linked ShotSpotter with a "reverse 911" notification system called the Communicator. When a gunshot is detected, the Communicator retrieves the phone numbers of nearby homes and businesses from a geographic database, calls them up, announces that the police are aware of the gunshot, and asks anyone who can describe the culprit to press a button and report what they know. Although the trial had yet to produce a sale, Microsoft chairman Bill Gates nominated ShotSpotter and the Communicator for a 2000 Computerworld Smithsonian Laureate Award, no doubt partly because they rely on Microsoft database and server software. The judges apparently agreed with Gates, and the two systems became part of the national museum's collection of "the year's most innovative applications of technology from 38 states and 21 countries." Clever indeed, but most police officials remain skeptical about either the need for a gunshot locator or its price.

+   +   +

Computer mapping is another matter. For more than a decade, police departments throughout the country have been mapping crimes and other "incidents" stored in their electronic databases. Although computers expedite the process, sticking pins on wall maps is an old strategy for guessing where a criminal might live or strike next. Law enforcement experts still refer to detailed maps of crime locations as "pin maps" even though mapping software marketed to police typically pinpoint crime locations with prominent dots (fig. 7.4), sometimes colored to show time of day or type of crime.

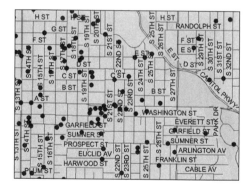

FIGURE 7.4 Portion of a pin map showing incidents of vandalism in Lincoln, Nebraska. The original color map is from City of Lincoln/Lancaster County, "Lincoln Police Department 2001 Crime Data," http://ims.ci.lincoln.ne.us/isa/2001Crime/.

Contemporary crime mapping is rooted in a long-held belief that law enforcement can benefit from mathematical analysis and operations research. The International Association of Chiefs of Police has encouraged systematic collection of crime statistics since the 1930s, various task forces of the President's Crime Commission promoted computer-aided data analysis in the 1960s, and isolated partnerships of university researchers and detectives explored applications of computer cartography in the 1970s. Although a few police departments made maps with computers in the 1980s, interest was sparse until the fortuitous convergence in the 1990s of specialized software, pedagogic propaganda, and federally funded demonstration projects.

Perhaps the greatest impetus was the Violent Crime Control and Law Enforcement Act of 1994, which promised to put 100,000 more cops on the street at a cost of $8.8 billion over six years as well as equip police agencies of all sizes with high-tech crime fighting tools, including geographic information systems. Federal grants helped many departments purchase a GIS, and the Department of Justice's Crime Mapping Research Center offered guidance in selecting and setting up a system and using its maps in daily operations. Established in 1997, the Center sponsors research on the spatial analysis of crime, maintains an information clearinghouse and online tutorial (www.ojp.usdoj.gov/cmrc/), and runs an annual conference for researchers and users. As part of its outreach mission, the Center published geographer Keith Harries's splendidly illustrated 204-page guide, *Mapping Crime: Principle and Practice,* and distributed more than 8,500 copies to targeted mailing lists that included crime analysts, police chiefs, and sheriffs.

Crime maps vary in appearance and purpose. Some mark the locations of individual crimes with point symbols like the dots in figure 7.4; some display counts by block, patrol zone, or district for a specific type of incident like auto theft or burglary; and some report counts as rates after dividing by each area's population, land area, miles of street or sidewalk, or number of households. While competent general-purpose mapping software offers all these options, only a GIS designed for crime analysis is likely to highlight hot spots. Defined as a small area with an abnormal number of crimes within a short period, a hot spot can be a street corner, a block, a schoolyard, or a neighborhood. Although many hot spots are persistent and predictable, a new hot spot might indicate a new burglar in the neighborhood, emerging conflict between street gangs, or a similar threat calling for prompt action. Mapping systems that automatically troll for similar crimes clustered in space and time can identify statisti-

Street Gang Turfs 1991
- Disciples
- Latin Gangs
- Vice Lords
- Disputed Turf
- Other Gangs
- Not a Gang Turf

● Homicide

Hot Spot Areas
▬ Turf Violence
— Drugs

FIGURE 7.5 Composite map relates homicide locations *(dots)* to hot spots for drugs *(thin ellipses)* and nonlethal violence *(thick ellipses)*. Based on 1987–90 data from the Chicago Police Department, the original color map is from Thomas F. Rich, "The Use of Computerized Mapping in Crime Control and Prevention Programs," *Research in Action* [National Institute of Justice newsletter], July 1995, 1–11; map on 4.

cally significant clusters and highlight their locations with ellipses, as in figure 7.5. The Chicago Police Department used this map to study territorial disputes between rival gangs. Note that turf violence and homicides are strongly correlated, whereas drug activity is generally less common along the perimeter of a gang's territory.

Maps can also fight crime by making precinct commanders answerable to district commanders and the chief of police. The most impressive example of map-based accountability is the ComStat (Computerized Statistics) program initiated by the New York City Police Department in 1994. Similar in principle to briefings at which battlefield commanders plan or critique a tactical assault,

ComStat also mimics the regularly scheduled meetings at which executives of large manufacturing and retailing firms review production and sales data. A similar flow of information occurs at police headquarters when precinct and district commanders gather twice a week for a three-hour early morning "crime strategy meeting." Each precinct commander makes a presentation once or twice a month using a large interactive map that displays timely crime patterns for incidents ranging from homicide to aggressive panhandling. Colleagues ask questions and offer suggestions, and higher-ups indicate their displeasure when a precinct's strategy doesn't work. In addition to underscoring effective tactics and emerging trends, the maps encourage collaboration in addressing hot spots along precinct borders as well as crime rings operating in different parts of the city.

ComStat is the brainchild of the late Jack Maple, a former New York transit policeman who demonstrated how mapping could cut down crime in subway stations. Later a private consultant, Maple helped sell ComStat to police departments in other large cities, including New Orleans and Philadelphia. Interviewed for *Government Technology*, he gave an insightful answer to the question "How does the mapping help?"

> The beauty of the mapping is that it poses the question, "Why?" What are the underlying causes of why there is a certain cluster of crime in a particular place? Is there a shopping center here? Is this why we have a lot of pickpockets and robberies? Is there a school here? Is that why we have a problem at three o'clock? Is there an abandoned house nearby? Is that why there is crack-dealing on the corner?
>
> By looking at this, you can figure out where you need to be and when. You can figure out what time the pickpockets are working. You can look at stolen cars—where they are being stolen from and where they are being recovered. If only the bones are being found, you know there is a chop shop nearby.
>
> A map can give you all this. Then you can start looking at patterns and chronic conditions.

As Maple's examples imply, ComStat maps encompass census and land-use data as well as police intelligence.

Law enforcement's ultimate accountability, of course, is to the public, which pays police salaries and elects city officials. Many departments recognize that citizens are not only their employers and clients but also potential partners in the "war on crime." And as partners, the public can benefit from timely information about when and where to be especially careful. Although police agencies typically guard their data, many have discovered Web cartography as an effective way to warn citizens about crime and recruit Neighborhood Watch volunteers. Unheard of a decade ago, police Web sites now have a key role in "community policing."

Not all police Web sites offer maps, and those that do vary widely in sophistication and flexibility. Static maps showing only beat or district boundaries are common, but a few agencies help visitors compose highly customized maps. An exemplar is the San Diego Police Department (www.sannet.gov/police/), which lets users se-

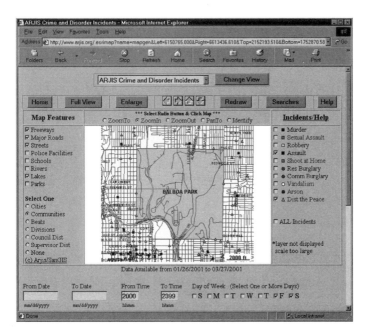

FIGURE 7.6 Among the neighborhoods surrounding San Diego's Balboa Park, the area to the southwest is especially prone to assault and rowdy behavior. The original color map is from San Diego Police Department, "Crime Statistics and Maps—Interactive Mapping Application," http://www.sannet.gov/police/stats/index.html.

lect from menus of crimes and reference features, specify starting and ending dates, and restrict the display to specific days of the week and times of day. Figure 7.6, the result of a request for a map of violent crimes reported near Balboa Park between 8 p.m. and midnight, covers a two-month period in early 2001. A visitor not satisfied with this view could zoom in or out, pan to nearby neighborhoods, or select other crimes and features.

I've heard critics charge that incident mapping ignores the underlying causes of crime. And they're right of course. Whatever the causes of crime—poverty, irresponsible parenting, inadequate education, impulsive behavior, and dysfunctional families and communities come to mind—maps will have a minor role, if any, in their abatement. But if we look instead at the causes of *victimhood,* mapping and other geographic technologies can improve the quality of life for almost everyone, and more so for the poor perhaps than for the rich. Moreover, as a *New York Times* editorial suggested in the wake of the Abner Louima lawsuit, ComStat can promote more effective law enforcement by including complaints against the police.

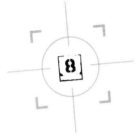

Keeping Track

Parents, do you know where your children are? What about the neighborhood pervert? In the 1980s TV spots highlighted the dangers of unsupervised teenagers out at night. Nowadays government-run Web sites spill the dirt on paroled sex offenders living in our neighborhoods. It's all there: names, addresses, offenses, and photos—a strong warning to keep our kids indoors at night and in plain sight throughout the day.

In addition to the "Megan's Law" Web sites that many states set up in the late 1990s to "out" sex offenders, GPS and radio triangulation can be used for tracking parolees and spouse abusers as well as for reporting criminal attacks, accidents, and medical emergencies. The instantaneous calculation and transmission of an individual's location is the ultimate double-edged sword: a technology offering creative alternatives to ostracism and incarceration while threatening cell-phone users, adolescents, and ambulatory consumers with an unprecedented loss of privacy. Although many applications have clear benefits with no apparent downside, the seemingly neutral label "location-based services" includes such neo-Orwellian scenarios as a digital leash based on an implanted chip that not only reports location but can administer pain if the subject steps outside a predefined perimeter—a human version of the "electronic fence" systems for dogs.

+   +   +

Megan's Law commemorates Megan Kanka, a seven-year-old New Jersey girl abducted, raped, and murdered in 1994 by a neighbor. Neither Megan's parents nor her neighbors knew that the three men sharing the house across the street were convicted sex offenders, and that one of them, Megan's killer, had spent six years in prison for sex crimes. Had the Kankas known of the occupants' criminal histories, victims' rights advocates argued, their daughter would still be alive. Within a week fifteen hundred people had signed a petition asking that every child molester's criminal record be revealed to anyone living within 1,000 feet.

Three months later the New Jersey legislature passed the Sex Offender Registration Act of 1994. Like the Washington State statute on which it was based, the new law sorted sex felons into three categories and required them to register with the local police upon release from prison. Those in Tier One, deemed unlikely to commit further crimes, were merely reported to their victims and other police departments in the vicinity. Those in Tier Two, with a moderate risk of reoffending, were reported to the principals of nearby schools, the directors of local women's shelters and day care centers, Boy Scout leaders, and persons in charge of other potential victims. And those in Tier Three, believed to pose a high risk to the community, were subject to a thorough outing by local police, who could publicize their presence with posters, flyers, newspaper ads, and television announcements.

Motivated by sympathy for young victims and the belief that community notification protects children from pedophiles, other states adopted similar statutes. And in May 1996 President Clinton signed a bill that amended the federal Violent Crime Control and Law Enforcement Act of 1994 by renaming it Megan's Law and requiring states to "release relevant information [about sex offenders] that is necessary to protect the public." Although Congress did not specify the form of notification, states that failed to comply would lose 10 percent of their federal law enforcement allotment.

Critics questioned the constitutionality and effectiveness of community notification. Civil libertarians condemned it as a violation of ex-convicts' right to privacy as well as an unconstitutional double punishment of persons who had served their time. Psychologists noted that most child molesters were friends or family members, not strangers, and mental-health professionals scrutinized risk-assessment systems that assigned offenders points based on their crimes and behavior. Police voiced concern about vigilantism after neighbors harassed several recently released offenders and vandalized their property, while legal experts challenged the statutes' wording and warned of a slippery slope leading to less privacy for everyone. Although judges struck down Megan's Laws in several states, including New Jersey, legislatures revised and rephrased their statutes to ensure constitutionality. By the end of the decade, all states had some form of community notification, often via the Internet.

It's hard to tell exactly how many states have online sex-offender registries: although the trend is definitely rising, the number keeps changing as states set up new Web sites or change their policies on public access. Because of considerable leeway in how states can comply with the federal community notification requirement, sex-offender Web sites vary widely in the type of information provided and who may see it. For example, a May 1999 survey by the Bureau of Justice Statistics turned up fifteen states with publicly searchable online registries; ten states with Web sites providing access only to law enforcement personnel or merely describing their registry, registration requirements, and notification policy; and five states with plans to set up a Web site. A June 1999 survey by law student Jane Small found sixteen publicly searchable online registries, including four that included low-risk as well as high-risk offenders. By contrast, a May 2001 visit to the list of state sex-offender registry Web sites maintained by the National Consortium for Justice Information and Statistics found URLs for thirty states with public online searching and for another six with online program descriptions. Because the list reported incorrectly that the District of Columbia "does not maintain an online sex offender registry"—the D.C. police had set one up two months earlier—I suspect additional states might have had one as well.

Curious about whether any (s)ex-cons might be living nearby, I sent my browser to the New York State Sex Offender Registry (criminaljustice.state.ny.us/nsor/index.htm). Although the full registry includes low-, moderate-, and high-risk offenders, only the "Subdirectory of High-Risk (Level 3) Sex Offenders" is publicly accessible online. Like New Jersey, New York uses a three-level classification system, which provides a legally acceptable rationale for protecting the public by ostracizing only high-risk offenders. By contrast, Connecticut's sex-offender Web site, which published the names, addresses, and photographs of *all* sex offenders, low risk as well as high risk, was found to violate offenders' right to "due process." In his April 2001 ruling, federal judge Robert Chatigny noted that the state went well beyond the federal mandate to protect the public from dangerous individuals.

Don't expect New York's online registry to include all high-risk offenders and provide up-to-the-minute reports on their whereabouts. A 647-word disclaimer warns that the subdirectory should be used cautiously. Wary of lawsuits arising from attacks by offenders who should be in the database but aren't, the Division of Criminal Justice Services, which maintains the Web site and "updates this information regularly," notes that entries might be incorrect or out of date. What's more, some arguably dangerous offenders are missing because a federal court ordered the exclusion of "Level 3 sex offenders who committed their crime prior to January 21, 1996 [when the state's Sex Offender Registration Act took effect] and were assigned to a risk level prior to January 1, 2000," when the state revised the list of crimes requiring registration. Because protection of the public does not trump the constitutional right to due process, the state can't brand an offender as high risk without an official evaluation by its board of examiners.

Equally wary of attacks on registered offenders or persons with a close resemblance, New York warns online users that "comparisons based on appearance may . . . be misleading" and that "anyone who uses this information to injure, harass, or commit a criminal act against any person may be subject to criminal prosecution." The effectiveness of these caveats is questionable, as is the requirement that users identify themselves before searching. Although entering one's name and address might deter some harassers, a savvy vigilante could access the registry at a public library by typing in a fictitious name and address.

Resisting the temptation to conceal my identity, I advanced to a screen that invited me to search for offenders by last name, county, ZIP code, or any combination. A search for Onondaga County returned a list of seventy-seven names, each underlined to indicate a link to a detailed record that included the registrant's name, address, physical description, photo, and a concise record of sex offenses (including for some the age and sex of the victim). As a caveat at the top of the list noted, some of the names were aliases, nicknames, or alternative spellings reported by offenders. Although some matches were obvious—David was also Rockin Dave, Ron was also Ronald, and Neal was Neil—inspection of all entries revealed a number of true aliases and whittled the list down to fifty-two unique entries, a mere hundredth of a percent of the county's 2000 population.

I can see why critics object to needless fears raised by online photos. Almost all fifty-two images looked scary, even the distorted black-and-white images that resembled botched scans of police mug shots. Context might well explain this reaction: the snapshots were obviously taken under conditions of duress, and we expect people who commit sleazy crimes to look sleazy. If I replaced these felons' headshots with candid photos of my colleagues, most of whom are exceptionally nice folks, you'd probably find their pictures equally repulsive. Adding to an unwholesome impression, the typical entry described scars or tattoos.

Hardly a surprise, the group was all male. Or at least I think so: one offender, who goes by "Chris" or "Christine," could have been a cross-dresser. The color photo suggested a hint of beard, which contradicted the red lipstick and dangling hoop earrings. The entry listed his or her sex as "unknown" and reported his or her crime as attempted sodomy with a sixteen-year-old male.

Information about an offender's crimes and victims could allay or sharpen the viewer's fear. For example, felons with multiple offenses seem to prefer one sex or the other as well as a particular age group. Parents of a seven-year-old girl thus might find it reassuring that the pervert next door prefers adolescent boys. Entries also list conditions of release for parolees and include a space for describing the offense or a modus operandi. Although the "explicit nature" of the latter seems to preclude publication on the Internet in almost all cases, a message invites interested citizens to consult the local police department or call the division's "for-fee 900 # Information Line."

Rightly or wrongly, a parent's or potential victim's greatest fear is the sex offender in the immediate neighborhood or along the way to the local playground. To help users zoom in quickly, the Web site allows a ZIP code search, which can narrow the list greatly—to zero in the case of 13224, where I live. But over in 13210, closer to the university, live three high-risk offenders, one not far from the food co-op where my wife used to take our daughter when she was young. He sexually abused an eight-year-old girl five years ago, the registry reveals, and he's got wheels: a 1986 Nissan Pulsar, which parents can warn their daughters never to go near. More astute parents and grandparents will no doubt call the state's 900 number. If the guy abused the child of a woman he was living with, say, he's probably far less a threat than a predator who attacked a stranger. I'd be leery of the bastard but better aware of why "naming and shaming" (as it's called in Britain) makes rehabilitation difficult for nonrecidivist sex offenders.

Among states using the Internet for community notification, New York reflects an intermediate position between Connecticut, which until recently largely ignored relative risk, and California, which uses its Web site to advertise a sex-offender listing stored on a CD-ROM and available for searching at the county sheriff's office or, in large cities, at police headquarters. Like New York's online registry, California's public-access CD-ROM contains only high-risk offenders. Entries include the offender's name and known aliases, age and sex, a photograph (usually) and physical description, names of the crimes that resulted in registration, and the county and ZIP code of last known residence—but not his exact street address. A prospective viewer must be over eighteen years old and able to "state a distinct purpose" for searching the registry. In addition to holding a valid California driver's license or identity card, the viewer must sign a statement affirming awareness of the purpose of the data and the illegality of using the information to "harass, discriminate or commit a crime against any registrant." As an added safeguard, the viewer must also assert that he or she is not a registered sex offender—unlike legislatures in most other states, California lawmakers were wary of pedophiles using a Megan's Law registry to contact each other and exchange snapshots of their victims.

Although California prohibits unregulated online viewing of its sex-offender registry and is stingy with geographic details, local police departments can inform the public about sex offenders living in their community in a more precise way, with maps. That's the approach in Fairfield, a city of 95,000 persons midway between San Francisco and Sacramento. A reflection of increased use of GIS in law enforcement, the police department Web site (www.ci.Fairfield .ca.us/police/) serves up online pin maps centered on each of thirty-one local schools and updated every three months. A menu lists the names of the city's four high schools, five middle schools, thirteen public elementary schools, and nine church-affiliated schools. Double-clicking on a name yields a customized map (fig. 8.1) showing a portion of the local street network with the school in question at the center of a brown circle a mile in diameter. Blue stars represent schools, and red dots mark the addresses of sex registrants, labeled by street name but not house number. Even though the dots are in the middle of the street, parents familiar with the area might have no trouble narrowing a location to a handful of residences.

Variation among the states in access to their sex-offender registries could provide a rich database for studying what works, what doesn't, and why. But a meaningful evaluation requires a clear sense of what we mean by *works* and how we might tease out the consequences, intended and otherwise, of diverse approaches. A key concern is that community notification of known and convicted sex offenders addresses only part of a wider problem. Although

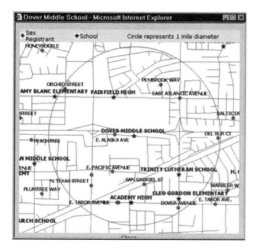

FIGURE 8.1 One of thirty-one school-centered sex-offender maps available from the Fairfield, California, Police Department Sex Offender Database, http://www.ci.fairfield.ca.us/police/map_list.asp.

Megan's Law Web sites can contribute to our children's wariness of strangers offering candy or eager to show off the new puppy in the back room, they do little to cut down on sexual abuse by Uncle Ralph, Father John, or Mom's new boyfriend. More troublesome is the need to balance our legal responsibilities to potential victims and known offenders. Although ostracism complicates an offender's return to a normal, productive life, it's clear that some individuals warrant continued surveillance once their sentences are served. Unless society and the courts are willing to incarcerate high-risk offenders for life or isolate them in "guarded villages," as social reformer Amitai Etzioni has suggested, community notification of some sort seems essential. How effective such notification is might well depend on the extent to which society accepts its responsibility for rehabilitation.

+   +   +

That community notification might be little more than a Band-Aid prompted some jurisdictions to experiment with a more proactive form of sex-offender surveillance: satellite tracking. An extension of the electronic ankle bracelets used to monitor parolees and pretrial defendants placed on home confinement, electronic tracking is a relatively inexpensive way to protect children from pedophiles and battered women from abusive husbands or boyfriends. And because vigilance is constant and discreet, the digital leash can also address the harmful effects of naming and shaming, the needs of parents who lack Internet access, and the reluctance of some offenders to report their addresses promptly and obey court-imposed restrictions.

Satellite tracking is a significant advance over older home-confinement systems based on a radio link between an ankle bracelet and a telephone set. A transmitter connected to a GPS receiver reports the offender's current position every 60 seconds to a surveillance center, where a GIS compares his location to a list of prohibited spaces such as schools, day care centers, strip clubs, bars, and the residences of the offender's victims or endangered spouse. In effect, the system establishes an "electronic fence," which can vary by time of day, for example, to let the offender go to work or return home along a prescribed route. If the subject moves outside his prescribed perimeter, the GIS can record the violation and alert authorities as well as send the offender a warning. More advanced systems designed to monitor victims as well as offenders can enforce an order of protection by repeatedly comparing a known stalker's location with that of his electronically protected prey.

The digital leash is not just for sex offenders and spouse abusers. Its steady panoptic gaze reduces the likelihood a parolee or person awaiting trial will commit a range of other crimes, petty or serious: by recording the subject's movements in both time and space, the computer produces an electronic trail, which can be correlated with crime locations to add or remove him or her, automatically, from a list of suspects. Not a substitute for a conscientious rehabilitation program, constant monitoring reduces the likelihood of missed appointments for personal counseling or drug treatment. It's no surprise, then, that cost-conscious corrections departments are experimenting with satellite tracking as an alternative to prison for a variety of offenses, especially nonviolent crimes like drug possession. Compared to incarceration, which can be more expensive than sending a son or daughter away to a good private college, the daily rental fee for a GPS-based tracking unit is cheap, typically less than fifteen dollars. What's more, taxpayers don't need to feed and clothe the gainfully employed detainee, who can then help cover the cost of monitoring or pay restitution to his or her victims. Given the option of electronic monitoring or satellite tracking, which is usually voluntary, prisoners rarely reject the chance to escape the confinement of a tiny cell.

For a "high-tech ball and chain" like the SMART (Satellite Monitoring and Remote Tracking) system, developed by Pro Tech Monitoring, the prime drawback seems to be the weight of the batteries. A 6-ounce transmitter strapped to the ankle communicates with a 3.5-pound Portable Tracking Device (PTD), which is about the size of two VHS tapes and can be strapped to the waist or carried in a knapsack or briefcase. The PTD contains a GPS receiver, a microprocessor, a radio receiver connected to the ankle bracelet, and a wireless modem linked to Pro Tech's surveillance center. The microprocessor notifies the center if the offender tries to tamper with the bracelet or violates the prescribed "inclusion" and "exclusion zones," which define where he or she should or should not be at various times and are stored on the PTD's computer. If the subject wanders out of bounds or gets too close to a victim or witness, the PTD emits a warning signal and after a few minutes reports the violation to Pro Tech, which in turn notifies the offender's "supervising agency" by pager, fax, or Internet. Zoning can also restrict a detainee to a particular state or county and declare off limits broad areas where terrain interferes with GPS or wireless reception. Because GPS does not work indoors, no-go zones might include large enclosed spaces like the Mall of America.

Although the offender must carry the PTD when traveling, it can be set down at home, work, or a friend's house. In this sense, the radio link between ankle transmitter and PTD is a programmable electronic leash. At night, when the PTD insists upon a comparatively strong signal from the transmitter, the leash is short and the offender must stay close by. At other times and locations the supervising agency can increase the allowed distance to as much as 1,000 feet so that the subject can play softball or work in a warehouse or garden. It's not a perfect arrangement because the detainee can always ditch the PTD, leaving a record of where he or she was, not where he or she is. But with smaller, lighter batteries and miniaturized components, developers can integrate the GPS, microprocessor, and modem with the ankle bracelet and reduce the chances that an offender who snaps can snap the leash.

It gets more Orwellian than Orwell's *1984*. "Third-generation" systems, talked about but yet to come, promise tamper-proof ankle bracelets that can monitor vital signs as well as location and can "shut down" a wayward offender who tries to remove the bracelet. And because of miniaturization, the units can be implanted beneath the skin like the contraceptive Norplant rather than merely strapped on. Max Winkler, a Colorado parole officer who extolled the benefits of third-generation monitoring in a July 1993 article in *The Futurist*, described a subdermal microcomputer with artificial intelligence software able to detect a dangerous pattern of vital signs and release a tranquilizer or soporific if a crime or unacceptable sexuality seems imminent. Although satellite surveillance has yet to catch up with Cyberpunk scenarios in which the "Autoinjector" and implanted "Poison Vial" offer workable solutions to twenty-third-century deviance, developers can easily add "punitive measures" to a location tracker worn as a belt or bracelet. In addition to phoning in the location of the wearer who dares turn a screw, a "monitoring and restraint system" can enforce exclusion zones by administering a mild, unobtrusive electric shock for a slight zone violation and upping the amps if the detainee fails to respond. After all, electric-shock restraints have been available for at least a decade to appearance-conscious dog owners who prefer an "invisible electronic fence" to the chain-link variety.

+   +   +

As an alternative to prison, satellite tracking seems a win-win strategy, with clear benefits for prisoners, parolees, and pretrial defendants as well as taxpayers. Assuming the technology performs perfectly and does not become a substitute for counseling and treatment, the principal threat is the ease with which society can broaden its notion of deviancy. It's a real threat, especially for an ostensibly "free" society whose history includes a willingness to experiment with Prohibition, condone restrictive covenants, and flirt with Political Correctness. Americans need not look to China and Iran for examples of well-intentioned repression.

A further risk lurks in satellite trackers designed to find lost children, wandering pets, and itinerant Alzheimer's patients—vulnerable subjects who need protection and are unlikely to resist intrusive monitoring. Parents, pet owners, and eldercare constitute a far larger market than the criminal justice system, and because peace of mind can be as valuable as protection, it's likely to prove a highly lucrative one. What's more, the spin-off potential is immense, especially for entrepreneurs who blend GPS with conventional technology or identify new uses of real-time tracking. For example, satellite systems designed to help trucking firms track equipment and manage employees can be reconfigured to help railroads avoid rear-end collisions. And the widely advertised OnStar automobile navigation system, which links GPS with a diagnostic computer and automatically summons help when a tire goes flat or an airbag deploys, might one day offer lower insurance rates to motorists willing to let an insurer monitor their driving.

Although privacy advocates are inherently leery of satellite tracking, no proposal has raised as many eyebrows as the Digital Angel under development by Applied Digital Solutions. In 1999, ADS bought rights to a patent granted two years earlier for an "apparatus for tracking and recovering humans [with] an implantable transceiver [designed] to remain implanted and functional for years without maintenance." Civil libertarians promptly warned of abuse by government snoops as well as cyber-savvy kidnappers on the lookout for lucrative prey. Critics included Susan Cutter, president of the Association of American Geographers, who condemned the "new locational e-slavery." Especially troubling to Cutter was the company's Web site (www.DigitalAngel.net), where a cartoon animation pictured an angelic winged figure swooping down to rescue Grandpa from cardiac arrest, repair motorist Jane's flat tire, and restore the bewildered Spot to his anxious owners. What loving son or daughter, loyal spouse, or concerned parent would not eagerly shell out $299 for a deal that includes a Web-delivered map (fig. 8.2) of the loved one's location, complete with street address, temperature, and pulse rate?

Offered in early 2001, the first-generation Digital Angels resembled wristwatches and belt-mounted pagers—a far cry from ADS's futuristic development plan. The firm's patent describes an "im-

Here's what the Digital Angel Delivery System looks like on a subscriber's Web-based computer:

As shown, the Digital Angel Delivery System can manage medical applications by gathering bio-readings such as pulse and temperature, and communicating the data, along with location information, to a ground station or call center.

FIGURE 8.2 The Digital Angel Corporation, an Applied Digital Solutions company, promises Web-accessible maps of the tracked person's location. Text and illustration from Digital Angel Corporation, "The Technology behind the Digital Angel," http://www.digitalangel.net/da/tech.asp.

plantable triggerable transmitting device" that conserves power by transmitting only when activated by the implantee or by a coded signal from the "tracking and locating center." Power is a key concern: although "an electromagnetic induction source . . . placed close to the body on a regular basis" can recharge the miniature storage battery, the preferred design relies on a power transducer to "derive power from physical work done by muscle fibers in the body." In times of stress "a novel sensation-feedback feature" activates the transceiver, and ground-based tracking expedites recovery. As a diagram (fig. 8.3) in the patent application shows, device D_1, implanted in the smiling, Gingerbread Man-like person P_1, can broadcast its position to antennas A_1, A_2, and A_3 or provide an electronic beacon for a recovery vehicle with mobile antennae MA and directional and mobile receivers DR and MR.

Jerry Dobson, a GIS expert at Oak Ridge National Laboratory and columnist for *GeoWorld,* objects vehemently to chip implants for tracking children who might be lost, kidnapped, or visiting a friend without permission. To dramatize the dangers of GPS-based "branding and stalking," he described a little girl's walk home from school. In an open field along the way a curious form, perhaps a rare flower or a small animal, attracts her attention. "Impulsively, she charges across the field. But suddenly, her biceps twitches. Before

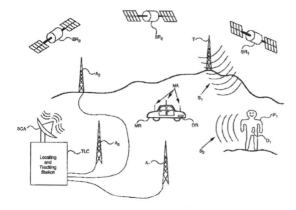

FIGURE 8.3 Figure 1 from the application for U.S. Patent no. 5,629,678, awarded May 13, 1997, and acquired by Applied Digital Solutions, describes key components of the inventor's plan for a "personal tracking and recovery system."

she can stop, her arm stings—then aches. She turns back, and her pain ceases at the sidewalk. Simultaneously, a commercial service provider reports to her parents." Although Dobson laments this lost "moment of discovery," he worries more about possessive parents reluctant to have the chip removed when the child grows up—could a "taking out" party become a new rite of passage?—as well as shrewd kidnappers not at all adverse to some ad hoc surgery.

Parental anxiety about their offspring's whereabouts is conspicuous in the growing number of teenagers with cell phones and pagers, used to keep in touch with friends but paid for by a Mom or Dad who appreciates the convenience of a wireless leash. And thanks to the Federal Communications Commission, parents will soon be able to track Junior's or Sis's location on a screen similar to the hypothetical Digital Angel Web map in figure 8.2 but for a lot less money. In 1996 the FCC, alarmed by a dramatic rise in 911 calls from cell-phone users unable to describe the location of a fire or accident, ordered wireless providers to pinpoint calls to within 125 meters (410 ft.). The original order let companies decide whether to insert a GPS chip in the handset or triangulate the caller's location with directional antennas on existing cellular towers. Removal of Selective Availability in May 2000 not only narrowed the reliability standard for handsets to 50 meters but jumpstarted a "location-based services" (LBS) industry, which is aggressively blurring the line between cell phones and PDAs. Forced to "geolocate" callers for emergency response, wireless carriers quickly discovered that the

10-meter accuracy possible with GPS could be a marketable commodity.

Loss of privacy is inevitable if the phone company can sell our coordinates to marketers and other stalkers. The wireless handheld that puts you at the center of its map or helps you locate the nearest hardware store is a marvelous invention, but who wants to stroll down Main Street and be bombarded with calls or e-mail from stores a couple doors away? Even more problematic than the commodification of location is the archive of electronic trails left by anyone talking on a cell phone or merely roaming. Bad enough that Big Brother and his mercantile siblings know where we are—without a restricted retention period and other safeguards they can easily discover where we've been.

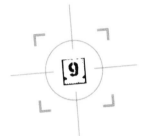

Addresses, Geocoding, and Dataveillance

For direct-mail advertisers and telemarketers, knowing where we live is nearly as useful as knowing what we might buy. Although few neighborhoods are perfectly homogeneous, most of us live near people similar in cultural conditioning, social aspirations, and spending preferences—or at least more similar than folks on the other side of town. And even demographically diverse districts can acquire an ambiance that attracts left-leaning political candidates and sellers of gourmet goodies and avant-garde clothing. Location is not a perfect predictor of consumer behavior or political preferences, but it works well enough most of the time so that people who have our addresses know—or think they know—a great deal about who we are.

How political strategists and other marketers exploit geographic proximity makes a fascinating if not frightening tale. Marketers, census takers, and locational consultants are its key characters, and software that matches addresses with neighborhoods and neighborhoods with buying habits is a recurring theme. Subplots include "neighborhood lifestyle segmentation" schemes for pigeonholing consumers by ZIP code and apportionment algorithms for integrating census information with data collected by postal code. Like much contemporary literature, the story has an ambiguous ending, inconclusive but insightful for readers who don't insist on simple solutions for complex problems.

+   +   +

What can you read into an address? Quite a bit, especially if it's on Fifth Avenue on New York's Upper East Side. Far less, though, if it's 302 Waldorf Parkway, my home in Syracuse, New York. The 302 tells you (correctly) that our next-door neighbors live at 300 and 304, while the Waldorf ambiguously suggests an expensively elegant Manhattan hotel, a popular brand of toilet paper, or a calorie-laden salad with apples and walnuts. More misleading is the appendage Parkway, which implies (incorrectly) two strips of pavement separated by a grassy median. Although planned that way, our street evolved as an ordinary town road with an oversize right-of-way. Equally confusing are the city and state parts of my address: we receive our mail through a Syracuse branch post office but pay taxes and vote in the Town of DeWitt.

If you're a catalog retailer, what really interests you is my ZIP code, 13224. According to recent estimates, our postal code includes 9,186 people living in 3,806 households. The median income of $51,125 is a cut above the national average of $41,914, but we're markedly more black (39.7 percent) and a shade less Asian (3.3 percent) than the country as a whole (at 12.4 and 3.9 percent, respectively). Not quite a melting pot but fertile territory for National Public Radio, Democratic politicians, and Lands' End outerwear. More interesting to most marketers are the area's "lifestyle clusters," which afford a concise description of not only my neighbors' consumer preferences but also their leisure pursuits, reading habits, and political inclinations.

Claritas Corporation, which pioneered clusters in the early 1970s, offers a free preview at its "You Are Where You Live" Web site (http://www.cluster2.claritas.com/YAWYL/). Two separate previews actually: one for PRIZM's "62 distinct lifestyle types, called clusters" and a second for MicroVision's "48 lifestyle types, called segments." An acronym for Potential Rating Index by Zip Markets, PRIZM evolved from a mere forty clusters in an earlier era of less diverse neighborhoods. Whereas PRIZM focuses on characterizing neighborhoods, MicroVision classifies individual households. Both approaches reflect the integration of census data with consumer and media surveys as well as diligently maintained records of catalog and online purchases. And while the ZIP code previews are free, both clustering strategies are available for a fee for much smaller neighborhoods defined by the Postal Service's nine-digit ZIP+4 codes, with between 5 and 15 houses on average, and the Census Bureau's block groups, with roughly 250 to 550 households apiece.

Although a five-digit ZIP code might contain as many as twenty different neighborhood types, the ClaritasExpress Web site reports only the top five, listed by number and name. For instance, in 13224, where I live, the most common clusters are

2	Winner's Circle
7	Money & Brains
8	Young Literati
19	New Empty Nests
30	Mid-City Mix

The numbers represent each cluster's rank on a 1-to-62 scale of affluence, while the names reflect a clever attempt to condense distinct lifestyles into glib yet meaningful labels. Although Winner's Circle might connote horse racing, not the rat race, a click on the name reveals a category of "executive suburban families" in the forty-five to sixty-four age group with professional occupations and well-above-average household incomes of $90,700. According to the Claritas databank, these households are likely to have a passport and a Keogh plan, shop at Ann Taylor, watch *NYPD Blue,* and read "epicurean" magazines like *Gourmet.* By contrast, the Mid-City Mix cluster consists of "African-American singles and families," either under eighteen or between twenty-five and thirty-four, in white-collar or service occupations and with household incomes of $35,000 and preferences for three-way calling, Pepsi Free, shopping at T.J. Maxx, watching *Nightline,* and reading *Muscle and Fitness.* These five diverse clusters and their interpretations confirm my hunch that 13224 is hardly homogeneous.

On a whim, I type in the ZIP code made famous by a TV show my daughter was addicted to back in high school, *Beverly Hills 90210.* Surprisingly, PRIZM identifies two groups prominent in 13224 as well as the country's most affluent cluster, Blue Blood Estates.

1	Blue Blood Estates
2	Winner's Circle
7	Money & Brains
10	Bohemian Mix

Only four clusters appear because the Web site doesn't include (at least not for free) clusters that account for less than 5 percent of a ZIP code's households. Fascinated by the apparent precision of these stereotypes, I key in 11104, for the Sunnyside section of Queens, New York, where my daughter lives now, on her own a year out of college. None of the five clusters is a perfect fit for Jo herself, but their names reflect the cityscape I recall from my last visit.

10	Bohemian Mix
27	Urban Achievers
29	Old Yankee Rows
45	Single City Blues
46	Hispanic Mix

Intrigued, I enter ZIP codes for the White House (20500), the Syracuse University campus (13244), Vandenberg Air Force Base, California (93737), and Pine Ridge, South Dakota (57770), an Indian reservation reputed to be the poorest place in the United States. Again, there are few surprises: the White House was "not found in the database," while the university and the air base yield single clusters aptly labeled Town and Gown and Military Quarters. And PRIZM seemed at least moderately on target for Pine Ridge, which is characterized by only two clusters, Agri-Business and Hard Scrabble, the latter typical of "families in poor, isolated areas" with household incomes of $18,100 and preferences for reading *True Story*, watching auto racing on TV, and using coupons to buy tobacco. Not a fertile market if you're hawking Godiva chocolates and Volvos.

Wholly absorbed, I looked at Claritas's other take on my Syracuse ZIP code. Despite different labels, the area's top five Micro-Vision segments are similar to their PRIZM counterparts in describing an eclectic mix of moving-up and getting-by:

8	Movers and Shakers
1	Upper Crust
24	City Ties
46	Difficult Times
45	Struggling Metro Mix

More significant is a richer array of preferences. Movers and Shakers, I learn, are "high income households containing singles and couples, age 35–49, with no children, one to two people." Yes, a few houses in my neighborhood match this description, but whether they "listen to National Public Radio [and] read *Golf Digest, Newsweek,* and *Car & Driver* magazines" is difficult to confirm. Equally puzzling is the revelation that the "low income young single adults, age 22–39, [in] one person households" in the Struggling Metro Mix segment favor "three-way custom calling" and "use [their] home PC more than 25 hours per week." Are they writers, I wonder? Or perhaps graduate students addicted to conference calls?

A competing neighborhood segmentation system named ACORN (for A Classification Of Residential Neighborhoods) presents a third but very different point of view based on forty-three clusters. Although the ACORN Web site offers a free ZIP code look-up, its reports reveal only the single most common segment—an unfortunate strategy for 13224, which the system assigns to segment 8E, Urban Working Families, summarized in four dismal sentences:

> Nearly 40 percent of this young group of single-parent families is under the age of 20. They are the working poor. They live in older, pre-war residential townhouse developments and small/multi-unit buildings. They buy take-out food, hair and skin-care products, baby products and children's clothing.

A few parts of the area might match this description, but they're hardly typical. And while my immediate neighborhood includes two single-parent families, no one would consider their middle-aged, professional moms exemplars of "the working poor." Retailers take clusters seriously, though, and ACORN's characterization might well explain the two Chinese take-out shops that opened recently in a nearby shopping strip.

Despite contradictions and ambiguities, lifestyle clusters conjured up by Claritas and other geosegmentation strategists not only choose targets for coupon promotions, fund-raising appeals, and political pitches but also help national chains locate new stores and restaurants. Marketing managers and development directors who rent or trade mailing lists consider screening by postal code or block group an effective way to increase yield. Each piece of mail is a drain on net take, after all, and a strategy that improves the response rate by just a few percentage points easily justifies the cost of screening. Political consultants are equally eager to identify swing voters, those not fully sold on their client or unlikely to vote without a reminder or a ride to the polls. Stakes are even higher for the retail or restaurant chain comparing potential sites or promoting a new location. Because the site that fails is embarrassing as well as costly, corporate planners appreciate GIS-based analyses that evaluate households within a given radius of a proposed site or send coupons only to addresses deemed receptive in both distance and lifestyle.

+ + +

How do Claritas and its competitors identify lifestyle clusters? Although proprietary algorithms preclude an exact reconstruction, we know that neighborhood lifestyle segmentation is a marriage of two types of data: census information tabulated for block groups, census tracts, and ZIP codes and consumer data based on purchases, consumer surveys, magazine subscriptions, and credit reports. Back in the 1960s geographers and sociologists used a form of cluster analysis to explore urban structure, but marketing executives had little use for this technique until Jonathan Robbin founded Claritas in 1971. According to journalist Michael Weiss, who made a career of mapping clusters and extolling their use, Robbin was a "Harvard-educated computer whiz" who saw potential value in converting census data, typically reported for "arcane units called 'tracts,'" into ZIP codes, which marketers could understand. To make the results useable, he reduced the hundreds of census variables like average rent, median years of education, and percentage of housing units with indoor plumbing to a mere thirty-four principal factors that, in the language of statisticians, "accounted for 87 percent of the variation among U.S. neighborhoods." After scoring every ZIP code on each of the thirty-four factors, a computer partitioned ZIP codes into forty clusters with generally similar factor scores. Robbin's choice of forty clusters was largely subjective: fifty clusters would have been more homogeneous and precise, on average, but more cumbersome as well, while some members of a thirty-cluster classification would have been overly vague. Following an enthusiastic response to the census-based Claritas Cluster System, introduced in 1974, Robbin integrated his groupings with survey, consumer, and media data and released PRIZM in 1978.

PRIZM's success is apparent in the emergence of competing systems like ACORN and MicroVision (which Claritas acquired through a merger in 1999) as well as the enthusiastic adoption of clustering in Europe, Canada, and South Africa. Wide acceptance is also apparent in the gargantuan effort devoted to revising characterizations yearly if not monthly, as neighborhood change outpaces the Census Bureau's ten-year update schedule. I've witnessed massive change on Waldorf Parkway in only three years, during which two elderly neighbors died, one moved to a warmer climate, and a fourth went into long-term care. The decennial census is still important, but credit bureaus and mailing-list vendors must aggressively track movers. They accomplish this tracking partly with the help of the U.S. Postal Service, which sells its monthly NCOA (National Change of Address) updates, based on those little cards we fill out when we move. And although Claritas and its competitors rarely revise their clusters—marketers appreciate stability—their counts and characterizations for small areas require constant monitoring and perceptive forecasting. According to geographer Jon Goss, marketing executives might not understand the "'black box' mechanics of the analysis," but they accept lifestyle segmentation anyway because cluster labels "fit with their own stereotypes" of American consumers.

Part of the enigma arises from the incompatibility of census and postal geographies. With a mandate to compile redistricting data, the Census Bureau focuses on the census block. Defined more broadly as "areas bounded on all sides by visible features, such as streets, roads, streams, railroad tracks, and by invisible boundaries, such as city, town, township, and county limits, property lines, and short, imaginary extensions of streets and roads," census blocks are easily delineated in neighborhoods with curved streets and cul-de-sacs. Blocks aggregate conveniently into block groups, with between six hundred and three thousand people, and block groups combine to form tracts, intended as generally homogeneous areas with between fifteen hundred and eight thousand residents. Because reliable intercensal rates of change require a stable geography, the bureau seldom adjusts boundaries except to subdivide burgeoning tracts with more than eight thousand people.

Easy to delineate, census blocks vary considerably in size and population. In the Waldorf Parkway area, for instance, the fourteen blocks in block group 4 of tract 147 range in population from 0 to 178. As figure 9.1 reveals, a census block can be much more than a simple rectangle. For example, block 4003, where I live, is not only bounded by parts of Waldorf and three other streets but includes both sides of the 100-block of Buffington Road. Other blocks are small and uninhabited. Block 4004, for instance, contains only by a small gasoline station, and block 4006 is a tiny public park. In rural areas with few roads, census blocks as large as several square miles are not uncommon.

By contrast, postal geographies focus on the street segments that bound a block. To reap the efficiency of mail presorted by nine-digit

FIGURE 9.1 Census blocks in block group 4, tract 147, Onondaga County, New York.

ZIP code, the U.S. Postal Service routinely assigns consecutive numbers to opposite sides of the street for residential neighborhoods like mine, where the letter carrier typically walks up one side of the block and down the other. What's more, in rural areas where carriers drive their routes and all mailboxes are on one side of the road, odd and even addresses share the same ZIP+4 code even though the Census Bureau assigns them to separate blocks. And as block 4003 demonstrates in figure 9.2, census blocks occasionally include more than one ZIP code. Although precision marketing firms would prefer that the Postal Service and the Census Bureau adopt identical geocodes, what works well for delivering mail is inappropriate for electoral boundaries, and vice versa. A dirty little secret of clustering is the imperfect match between data collected by census tract and consumer statistics tabulated by ZIP code.

Not to worry. Cluster experts learned to apportion census tract counts among postal areas by assuming, for instance, that the ZIP code with 10 percent of a tract's population will contain 10 percent of the tract's foreign-born college graduates. This assumption is crucial because detailed demographic statistics based on the "long-form" questionnaires filled out by only one in six households are not available at the block and block-group levels. Apportionment begins with a block-by-block assignment of households and population. A computer calculates geographic center points for all census blocks, compares these centers with postal boundaries, and assigns each block to the ZIP code that contains its center. The software then adds the block's population and household counts to the respective totals for its dominant ZIP code (13224 in the case of block 4003 in figure 9.2). Housing and socioeconomic data available only at the tract level can then be apportioned according to population or households. If the center-point approximation assigned 10 percent of a census tract's population to ZIP code 13224, for instance, and if the tract contains two hundred foreign-born college graduates, the computer assigns twenty of them to 13224. Although proportional allocation seems reasonable, most or all of the tract's two hundred foreign-born graduates could live in the 13224 ZIP code.

The Census Bureau has its own strategy. Although the agency never discloses data for individual households, it will produce special tabulations for a fee. ZIP code aggregation began in 1972, when the bureau released a partial count by postal code. In 1981, after the Reagan Administration's budget cuts killed plans for a more complete publicly funded ZIP code tabulation, a group of ten marketing firms offered to pay $25,000 each for a detailed postal-zone count of the 1980 Census. In what one firm's president called "the sweetheart deal of the century," census officials accepted the proposal and withheld the results from other users for eighteen months. ZIP

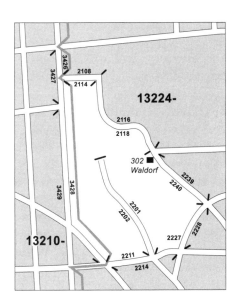

FIGURE 9.2 ZIP+4 postal codes for block 4003 and surrounding streets. The four-digit numbers are extensions of 13210 and 13224, the ZIP codes to the left and right, respectively, of the gray line.

Extending ZCTA Coverage

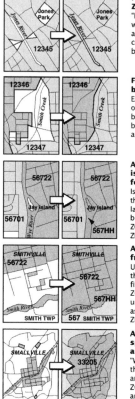

Fill holes within a ZCTA coverage
"Holes" (census blocks with no addresses) are assigned the ZCTA code of surrounding blocks.

Fill gaps between ZCTAs
Each ZCTA grows outward to fill adjacent blocks until all gaps between ZCTAs are assigned a ZCTA code.

Assign ZCTAs to islands and water features
Islands are filled with the ZCTA code of nearby land blocks. Water blocks get a special ZCTA code (three-digit ZIP Code + "HH").

Assign ZCTAs to fringe areas
Unassigned blocks on the edge of a ZCTA are filled with the adjacent ZCTA code. Large undeveloped areas are assigned the three-digit ZIP Code.

Assign ZCTAs to sparsely settled areas
"Windows" that include the ZCTA "fragments" are assigned to one ZCTA. The rest of the area is assigned the three-digit ZIP Code.

FIGURE 9.3 Because the U.S. Postal Service largely ignores areas not requiring mail delivery, the Census Bureau must fill holes and gaps in ZCTA coverage. From U.S. Census Bureau, *ZCTAs: ZIP Code Tabulation Areas for Census Data Products* (1999), 4.

code counts are now a standard census product, and in 2001, the bureau released boundary files for ZCTAs (ZIP Code Tabulation Areas), designed to help users map postal tabulations for Census 2000. By assigning each census block to the ZIP code with the most addresses and filling in gaps between ZCTAs (fig. 9.3), census officials created an expedient generalization of postal geography. It's an approximation, obviously, but so are most census products.

Although some marketing firms compile household data as well as tabulate by ZIP+4 areas, official census information summarized by five-digit postal code is useful as a reality check. What's more, geodemographics consultants can take advantage of five-digit

ZIP code tabulations by the Internal Revenue Service. Although some taxpayers might not be totally honest with the IRS, locally detailed tabulations for total income, salaries and wages, taxable interest, earned income credit, adjusted gross income, number of personal exemptions, and the numbers of taxpayers using schedules C (business income) and F (farm income) not only complement census data on income but reflect population shifts officially counted only once every ten years.

+ + +

If you think the federal government is Big Brother, guess again. Firms called "data warehouses" have gone well beyond the Census Bureau in not only collecting detailed information on individual households but also renting it to retailers, insurers, and even detective agencies. Acxiom, a data vendor with what Consumers Union calls the "largest collection of U.S. consumer, business, and telephone data available in one source," offers telemarketers and other clients a wide variety of personal information gleaned from credit reports, public records, consumer surveys, and credit card transactions. Data warehousing also includes niche firms like Moving Targets, a self-described "new resident direct marketing" firm that rents lists of "just-moved-in families" carefully selected from the Postal Service's National Change of Address (NCOA) data. And, as we have seen, Claritas clients find lifestyle clusters useful at the household level as well as for ZIP codes and census tracts.

What's scary is the ease with which data collected for one purpose, such as motor-vehicle registration, can be linked to information compiled for other purposes, such as consumer loans, voter registration, or health insurance. And it's not sufficient for government to restrict the use of our Social Security numbers. Gary Marx, a sociologist concerned with privacy issues, notes that an experienced snoop can dig up dirt or ferret out personal details by using a person's name, address, and date of birth to link records in different databases. Home telephone numbers can be especially useful as "consumer tags," and electronic credit card transactions afford nosy marketers further insights by pinpointing our shopping transactions in time as well as space. In addition, address-matching software makes it easy to find a household's census tract and block as well as estimate latitude and longitude, which are useful in plotting maps, calculating distances, and making inferences about personal indiscretions. For example, recurring room charges at motels within fifty miles of home might suggest a subject is having an extramarital affair.

Web browsing is also under surveillance thanks to software that can link our home and Internet addresses whenever we order online, fill out an electronic survey, or respond to an e-mail offer. "Geotargeting," a technology for linking Internet addresses to geographic locations, lets Web ads tout local firms. Web merchants and information vendors know a lot about our interests because of small files called cookies, which they place on viewers' hard drives, ostensibly to help them identify return visitors. Cookies let Amazon.com recommend new books, CDs, and DVDs similar to those we've purchased and help advertising firms like DoubleClick customize the banner ads that litter commercial Web pages. Thanks to cookie-based profiling, a viewer who frequently visits gardening Web sites is likely to see garden-related ads when visiting more general Web sites like *CNN.com* and weather.com. (And it might well explain the numerous offers of miniature video cameras that started popping up on my screen a few days after an intensive online search for information about web cams.) Largely benign, cookies can reveal preferences we'd rather remain hidden as well as encourage the White House Drug Office, which paid DoubleClick to track use of its Web site (www.whitehousedrugpolicy.gov), to confuse curiosity with intent. More invidious is the possibility of a "cache attack" by devious Web firms that can find out what Web sites we visited recently by scanning the cache of files our browsers store temporarily to expedite display.

Computer scientist Roger Clarke coined the term *dataveillance* to describe the "systematic use of personal data systems in the investigation or monitoring of [people's] actions or communications." Dangers include witch-hunts and illegal blacklisting as well as the harmful consequences of misidentification and erroneous data. In a capitalist milieu, in which sales and profits upstage sinister totalitarian scenarios, geodemographics firms and their clients use dataveillance to manage as well as stimulate consumption. Marketing models tell firms not only what to produce but where and how to sell it and to whom. Geographer Stephen Graham, a critic of GIS-based behavior modeling, sees this "surveillant simulation" as a force for increased segregation and polarization.

Is geographic dataveillance affecting the way society constructs places? Definitely, argues GIS scholar Michael Curry, who claims that widespread use of integrated systems fosters the impression that invasive profiling of individuals and places is both inevitable and beneficial. Conditioned to accept dataveillance as natural, most consumers consider resistance difficult if not futile. And when clusters and junk mail contribute to a sense of identity and well being, some of us even find the attention flattering. Because clusters reinforce the social status of an address, they explain the bitter complaints of people placed in a less classy ZIP code by a readjustment of Postal Service boundaries.

More serious are the dangers to personal privacy and the dilemma of too little or too much regulation. Harlan Onsrud, a legal scholar interested in the societal impacts of GIS, observed that highly local geographic data and spatial technology's prowess in integrating diverse databases foster invasions of privacy. Although he would prefer not to have information on individuals in a GIS, Onsrud recognizes competitiveness as a justification. Favoring self-regulation, accountability, and openness, he believes strongly that individuals should have the right to opt out. As a minimum, GIS managers should know the sources and reliability of their data and should be wary of the hazards of mixing data collected for different purposes or at different times. But government regulation could be stifling, Onsrud warns, especially if lawmakers use privacy protection as an excuse for restricting access to public information. Self-regulation might not be practicable either because "what is agreed to be 'smart business practices' by a large majority of practicing professionals may be considered highly unethical" by most citizens. Complex issues, conflicting mores, and evolving technology suggest that conflicts concerning data privacy might never be satisfactorily resolved.

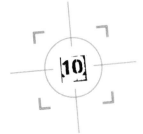

Case Clusters and Terrorist Threats

Mapping, GIS, and geographic modeling play key roles in the surveillance and control of disease. In identifying, monitoring, forecasting, and controlling epidemics, public health officials at the local, state, national, and international levels depend upon rapid and efficient collection and display of highly personal data. Health agencies also compile statistics on noninfectious diseases like cancer, many forms of which are poorly understood, and in doing so, confront conflicting requirements of confidentiality and public access. Although some disease maps are little more than descriptive propaganda or political palliatives, geographic information helps in tracking epidemics, predicting impact, distributing vaccines, spraying insecticide, administering quarantine regulations, evaluating the effectiveness of prevention and prophylactic campaigns, and communicating with medical personnel, emergency management officials, and the general public. And as I note at the end of the chapter, threats of biological warfare and bioterrorism pose new challenges for dealing with highly contagious diseases like smallpox as well as deadly pathogens manufactured in a laboratory.

+ + +

If disease mapping has a poster child, it's John Snow (1813–1858), the London anesthesiologist credited with discovering the waterborne transmission of cholera. Among geographers, Snow is best known for his 1854 map showing victims' homes clustered around Soho's infamous Broad Street Pump, which he identified as a source of contaminated water. According to epidemiological lore, the good doctor tried unsuccessfully to convince public officials to close down the pump. Undaunted, he took matters into his own hands, removed the pump's handle, and demonstrated the correctness of his theory when new cases plummeted. Truth be told, the epidemic had already run its course. What's more, Snow made his famous dot map several months later, for a revised edition of his book on cholera transmission. Even so, his pin map continues to embellish discussions of GIS and disease.

Medical geographers, GIS experts, and some epidemiologists perpetuate the Snow myth because it promotes disease mapping as a discovery tool and enhances the stature of their own disciplines. But a careful examination of Snow's writings indicates that he understood cholera's mode of transmission well before he made the map. Moreover, contemporary investigators with a different sense of the disease's origin and transmission produced more accurate maps of the Soho outbreak but misread them as evidence that foul air, not leaky cesspools, had spread the disease. Although Snow was a thoughtful observer, neither his map nor those of his rivals were of any value in generating insightful hypotheses. Snow's famous cholera map was pure propaganda—and copycat propaganda at that—but proved eminently useful later in the century, when public officials needed convincing arguments to isolate drinking water from sewage.

Belief in the power of visualization accounts for numerous atlases describing mortality variations among counties or larger units like state economic areas and hospital service areas. In 1975 the U.S. Public Health Service published its first disease atlas with an introduction that asserted "geographic patterns of cancer are useful in developing and testing etiologic hypotheses." Convinced that spatial patterns, if competently displayed, will reveal causal connections, the authors suggested that "perhaps the greatest value of the maps will be to designate high-risk communities where analytical-epidemiologic studies may detect specific carcinogenic hazards"— as if a ranked list of counties might not more efficiently point out the highest rates. A 1996 atlas praised this premier effort for revealing two "previously unnoticed clusters of high-rate counties" for which field studies "uncovered . . . the links between shipyard asbestos exposure and lung cancer and [between] snuff dipping and oral cancer." But in neither of the "field studies" cited did investigators rely on further mapping, and in neither instance were the suspected causal factors previously unknown. Although the 1975 cancer atlas might have fingered some intriguing hot spots, there is little evidence its maps ever "generat[ed] etiological hypotheses."

Still, the belief persists that maps will discover a cause, indict a polluter, or suggest a prophylaxis. And while it's easy to dismiss this cartographic scrutiny as misguided, maps produced for questionable epidemiological studies of complex, multifactor diseases no doubt promote public awareness, early detection, and lifesaving treatment. That's my assessment of the Onondaga County Breast Cancer Mapping Project, a local effort that registered an impressive 30-percent return for questionnaires included with residents' monthly mailing of Valpak coupons. The short survey, which my wife and daughter dutifully filled out in early 1998, asked for the respondent's home address, length of residence, race, date of birth, type of health insurance coverage, and the dates of most recent mammogram and breast examination by a health-care professional. Current and former breast cancer patients were asked to indicate when the tumor was discovered, the hospital at which they were treated, and where they were living at the time. Useful details, perhaps, but unlikely to reveal a normal American's exposure to carcinogens on the job or at a former residence.

Although our county executive announced that local breast cancer rates were not out of line with rates elsewhere in New York, survey participants have yet to see a summary of the data. Our only evidence of nonexceptional mortality is a ZIP code-level breast cancer map posted in April 2000 on a state health department Web site, which includes similar maps for lung cancer and colorectal cancer. Based on the state's tumor registry, which is more reliable than a mail-out survey, the Web map for Onondaga County revealed no apparent connection to race, ethnicity, income, or any other known factor for a disease especially common among affluent white women with Jewish ancestors from eastern or central Europe. Most mortality maps are equally baffling if not reassuring. Voters who don't find the absence of a clear pattern comforting might at least appreciate the concern of public officials willing to collect and map the data. Because such a map is a sign that women's health matters, breast cancer projects sometimes stimulate pleas from men for cartographic studies of prostate cancer.

Critics attacked the state's maps as not only vague but unresponsive to Long Island, which has some of the highest breast cancer rates in the nation. According to Richard Brodsky, chairman of the State Assembly's Environmental Conservation Committee, the maps "don't really identify cancer clusters, which occur at the neighborhood level, well below the ZIP Code level, and they haven't told us anything about environmental factors." Health officials, Brodsky charged, had more detailed data but weren't analyzing them. Echoing his complaint that "it's a sin to have this information and withhold it," Debbie Basile of the Babylon Breast Cancer Coalition commented, "it's a first step, but I want to see more."

While relatively detailed maps occasionally reveal suspicious clusters, careful examination typically ferrets out a nonenvironmental explanation. For example, address-level data collected by volunteers for the West Islip Breast Cancer Coalition exhibited a vague clustering of breast cancer cases in the west central portion of this affluent community on the south shore of Long Island. Especially troubling was the cluster's appearance on a map of women who had lived in the same house for at least thirty years, sufficiently long to share a common exposure. Discernible but not blatant, the cluster lies between the labels for Sunrise Highway and Union Boulevard on figure 10.1, which portrays cancers cases with large dots and other

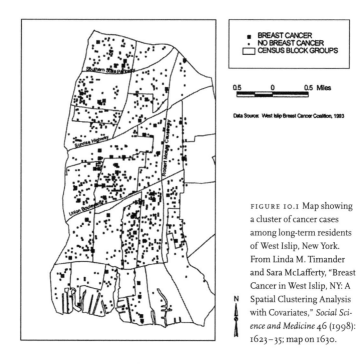

FIGURE 10.1 Map showing a cluster of cancer cases among long-term residents of West Islip, New York. From Linda M. Timander and Sara McLafferty, "Breast Cancer in West Islip, NY: A Spatial Clustering Analysis with Covariates," *Social Science and Medicine* 46 (1998): 1623–35; map on 1630.

women in the risk pool with smaller dots. Geographers Linda Timander and Sara McLafferty, who used statistical cluster-detection software to compare the spatial patterns of cancer cases and all long-time female residents, confirmed that the clustering of cases in this part of West Islip was hardly random. But a closer look at the cluster attributed most of the cases to one of two known risk factors: a family history of breast cancer and first pregnancy at age thirty or older. Once these correlates were factored in, the cluster disappeared. As the researchers noted, "although the finding of no spatial clustering may be unsatisfying to local residents, it does offer a constructive response to their questions."

State officials had similarly detailed data—but more complete and reliable data, I'm sure—and as Assemblyman Brodsky charged, they were holding back. For a sound reason, though: confidentiality. To gain the cooperation of patients and physicians, addresses as well as names of cancer patients are not released as either lists or dot maps. Nor is similar information provided for a host of illnesses, principally communicable diseases like HIV infection, syphilis, and gonorrhea, which physicians must report to their state or local health department. Notification, as it is called, is essential if public health officials are to monitor and control communicable diseases as well as investigate medical enigmas like breast cancer.

Grassroots groups like the West Islip Breast Cancer Coalition, which collected the data for figure 10.1, are not similarly constrained. Educated women, who do not consider cancer a stigma, lobby vigorously for detailed environmental investigations and often launch their own studies when the public health bureaucracy is slow to act. Committed to openness as well as scientifically curious, health activists readily reveal spatial details that public officials are compelled to conceal.

By contrast, health departments, which typically preserve anonymity by analyzing and reporting disease data by ZIP code or census tract, rarely release maps with the spatial specificity of figure 10.1. And if they do, they probably shouldn't: precisely positioned dots and roads are simply too revealing. A "map hacker" could register street intersections to a precise geographic framework, convert dot centers to grid coordinates, and use "reverse address matching" to recover household addresses, which could then be sold to mail-order sellers of unproven cancer cures or to enterprising lawyers recruiting plaintiffs for class-action lawsuits. Researchers can protect privacy by displacing each dot a random distance in a random direction—just enough to thwart unscrupulous snoops—and warning readers of the necessary subterfuge. Although "geographic masking" makes the maps slightly less authentic, investigators can still probe the unperturbed data for statistically significant case clusters—one less excuse for health officials to keep citizens in the dark.

+   +   +

Not all disease maps are visual placebos. Cartographic analysis contributes to eradication and remediation campaigns and, as John Snow's cholera map demonstrates, to public information efforts as well. When the health department decides to contain West Nile virus, for instance, citizens and local officials will need to know where infected birds were found and which neighborhoods require spraying. And police, highway, and health departments that request funds for traffic-accident prevention need to know where accident rates are especially high as well as related details that might indicate a need for road improvements, better signage, traffic lights, lower speed limits, or additional traffic cops. Clusters of child pedestrian accidents, for instance, can pinpoint areas requiring more vigilant enforcement of traffic laws in the late afternoon and information campaigns for parents of children attending nearby schools and playgrounds. Traffic accidents are very much a public health problem, and geographic information systems can be especially useful in developing prevention strategies and evaluating their effectiveness.

GIS can extract added value from data routinely compiled for disease surveillance. In Baltimore County, Maryland, for instance, public health professionals working with a GIS specialist identified high-risk areas for Lyme disease, a debilitating illness transmitted by deer ticks. Address-matching software converted the addresses of Lyme patients (cases) and a sample of randomly chosen households (controls) to point coordinates, which were then related to map overlays for fifty-seven separate environmental variables. Eleven of these factors proved statistically significant and provided the basis for a countywide risk map, which demonstrated the disease's strong association with steeper slopes and proximity to forested land. The researchers validated the map with data for the following year but declined to elaborate on the map's role in intervention. Medical geographer Ellen Cromley and her co-workers, who developed a GIS-based model of high-risk areas for Lyme disease in south central Connecticut, were less inhibited. Risk mapping, they argued, is useful for long-term strategies like discouraging low-density residential development in rural settings as well as for short-term tick-control measures like spraying acaricides and removing yard waste.

I found another example closer to home. My colleague Dan Griffith was the lead GIS investigator in a collaborative study of elevated blood-lead levels among children in Syracuse. The county health department, which had been screening children for lead poisoning since the early 1970s, was a rich source of data that had never been properly analyzed. Using parallel explorations of point, block, block-group, and census-tract data, Dan's team detected two parallel "swaths" with above-average blood lead, which matched patterns of older housing more closely than the configuration of heavily trafficked streets. Soil samples confirmed that lead concentration was generally higher around older dwellings, with wood siding and multiple layers of lead-based paint, than near major roads, with greater past exposure to lead additives in gasoline. Because "housing quality and maintenance practices" were more relevant than "minority status," the researchers recommended that intervention activities be focused on impoverished neighborhoods with older housing stock. Further research might recognize historical geography as important as soil-contamination data.

+ + +

On the opposite side of the confidentiality divide between health departments and other local officials is a mélange of information on technological hazards and ecological vulnerability. Environmental health units monitor air and water quality, environmental protection agencies compile annual reports on industrial releases of toxic substances, and fire departments and emergency management officials track shipments of hazardous materials as well as inventory local caches of explosive or toxic chemicals that warrant evacuation plans and other precautions. Essential to land-use planning and code enforcement, these data have been generally accessible— at least before September 11, 2001—to individuals and citizens groups as a result of "right-to-know" laws.

Making sense of the data is another matter. The U.S. Environmental Protection Agency, which collects annual reports on industrial emissions and discharges for its Toxics Release Inventory and publishes the information on the Internet (www.epa.gov/tri), makes little effort to help citizens assess impacts on lungs or livers. At a companion Web site (www.epa.gov/enviro/html/em/), the EnviroMapper pinpoints release locations and toxic-waste sites at various scales but does not integrate this information with maps of pollutants in the air, water, or soil. Arguably useful in keeping the spotlight on individual polluters and specific toxins, the Enviro-Mapper has only the weakest link to human health. A more complete picture of a contaminant's impact on drinking water, for instance, might require access to a GIS that integrates chemical and bacteriological analyses for test wells with confidential illness data and perhaps a computer model of local aquifers. But even if access were permitted, few citizens would know how to proceed. Socially conscious GIS researchers have proposed the development of "public participation geographic information systems." Their goal is twofold: helping citizens understand the limitations and efficient use of spatial data and involving them more fully in community decision making. Although user-friendly software and broader access to data are desirable, whether PPGIS becomes an instrument of empowerment rather than marginalization depends largely on the formation of partnerships between grassroots organizations and sympathetic GIS experts. A promising example is the Community Mapping Assistance Project (CMAP) initiated by the New York Public Interest Research Group in 1997. In its first three years, CMAP staff assisted a hundred grassroots groups and other nonprofit organizations with more than 250 projects. Although most assignments addressed community planning or social services, CMAP produced a number of compelling cartographic posters like figure 10.2, in which an alarming disparity in pediatric lead levels in Brooklyn demands action.

+   +   +

For health officials charged with spotting and suppressing highly contagious diseases among people or animals, mapping is more than a propaganda tool. Geographic data systems play an important role in public health management not only by expediting the collection, organization, validation, and exchange of epidemiological data but also by promoting proactive studies of local and regional health care. Examples include tracking the spread of raccoon rabies across a region—field investigators with GPS units can easily record the coordinates and condition of infected animals—and using the maps to plan and publicize rabies clinics for household pets.

Spatial analysis helps local health agencies cope with a variety of environmental threats. A typical application is the routine monitoring of household septic tanks near municipal water wells. Because drinking water is vulnerable to contamination from malfunctioning septic systems, health officials can prioritize field inspections by defining small circular buffers around all septic tanks listed in the environmental permits database and overlaying these threat zones

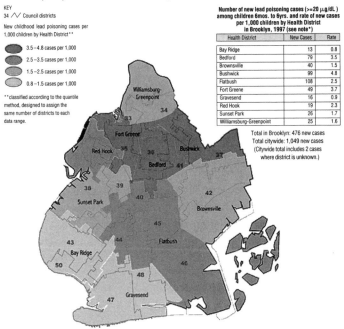

**New Cases of Childhood Lead Poisoning (>=20 μg/dL)
per 1,000 children 6mos.–6yrs. of age by Health District in Brooklyn, 1997***
(comparative statistics showing relative rates within Brooklyn)

KEY
34 /\/ Council districts

New childhood lead poisoning cases per
1,000 children by Health District**

- 3.5–4.8 cases per 1,000
- 2.5–3.5 cases per 1,000
- 1.5–2.5 cases per 1,000
- 0.8–1.5 cases per 1,000

**classified according to the quantile
method, designed to assign the
same number of districts to each
data range.

Number of new lead poisoning cases (>=20 μg/dL)
among children 6mos. to 6yrs. and rate of new cases
per 1,000 children by Health District
in Brooklyn, 1997 (see note*)

Health District	New Cases	Rate
Bay Ridge	13	0.8
Bedford	79	3.5
Brownsville	40	1.5
Bushwick	99	4.8
Flatbush	108	2.5
Fort Greene	49	3.7
Gravesend	16	0.9
Red Hook	19	2.3
Sunset Park	26	1.7
Williamsburg-Greenpoint	25	1.6

Total in Brooklyn: 476 new cases
Total citywide: 1,049 new cases
(Citywide total includes 2 cases
where district is unknown.)

FIGURE 10.2 Pediatric lead poisoning in Brooklyn as mapped by the New York Public Interest Research Group's Community Mapping Assistance Project. Original color map from the CMAP map gallery, http://www.cmap.nypirg.org/map_gallery/default.asp.

on a map of primary aquifers and a map of concentric proximity zones around water-supply wells (fig. 10.3). Records of past violations as well as a hydrologic model describing the flow of ground water provide an even more precise identification of septic systems requiring more frequent inspection. Ideally the GIS would also include underground gasoline storage tanks and chemical plants as well as the locations of monitoring wells, drilled specifically for testing water quality.

Modeling can also inform cancer-screening programs, as demonstrated by government efforts to identify possible victims of thyroid cancer in eastern Washington. Between 1945 and 1951 a plutonium plant near Hanford, Washington, released substantial

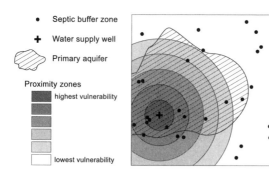

- Septic buffer zone
- + Water supply well
- Primary aquifer

Proximity zones
- highest vulnerability

- lowest vulnerability

FIGURE 10.3 Schematic example of an overlay analysis for prioritizing inspections of septic systems that might threaten local drinking water.

amounts of airborne radiation, which prevailing winds dispersed eastward toward farms at which pasture-fed dairy cattle contaminated locally marketed milk with Iodine-131. Because the thyroid gland concentrates radioactive iodine from the blood, and children typically consume more milk than adults and infants, anyone who grew up in the area during the late 1940s should be evaluated for thyroid cancer. To help identify candidates for screening, researchers at the Centers for Disease Control estimated the radiation dose for average milk-drinkers in 1945 by linking a GIS and an atmospheric circulation model. Health officials then invited persons who had lived in the area to request an individual dose assessment based on their exact residential histories.

Health departments can actively combat illness by asking hospitals, health professionals, and social service agencies to collect data on infants and other vulnerable populations. In North Dakota, for example, state health officials compared birth rates with visits for prenatal care in order to identify areas requiring better health education or improved access to health services. For Hillsborough County, Florida, state officials analyzed survey data on the vaccination of two-year-olds, mapped vaccination rates by block group, and detected "immunization pockets of need," in which outbreaks of childhood diseases were especially likely. In DeKalb County, Georgia, the board of health used PRIZM lifestyle clusters to target block groups in minority neighborhoods for antismoking campaigns.

GIS also helps health officials investigate eruptions of contagious diseases. For example, when an outbreak of infectious diarrhea struck soldiers and their families on a North Carolina military base, maps of patients' quarters focused attention on a specific housing area. Field interviews of residents identified a plausible means of interfamily transmission: small wading pools set up outside for children, some of whom might have contaminated the water with fecal matter. In a replay of the Snow myth, banning the pools abruptly halted the epidemic.

In controlling malaria and similar diseases transmitted by mosquitoes or other insect vectors, health officials integrate GIS with remotely sensed imagery to detect and monitor breeding areas. As a supplement to meteorological instrumentation and field sampling, color-infrared aerial photography and high-resolution satellite imagery can describe the extent of swamps and other habitats as well as their proximity to humans and livestock. Spatial modeling, another component of vector surveillance, helps officials predict outbreaks, assess relative risk, and orchestrate carefully timed spraying of vector habitats.

Animal and plant diseases, which can infect humans and devastate agriculture, warrant use of spatial analysis for epidemiological surveillance and decision making. Whether the disease is as episodic as raccoon rabies or as sporadic as foot-and-mouth disease, GIS can be a useful instrument for early identification as well as a powerful tool for distributing vaccine and enforcing quarantine. And when an epizootic starts spreading rapidly, like the European outbreak of foot-and-mouth disease in early 2001, GIS can help officials orchestrate containment by identifying entry points, suggesting likely paths of transmission, and delineating areas in which movements of people, vehicles, and agricultural products are restricted. During the 2001 episode, for instance, Britain's Ministry of Agriculture, Fisheries, and Food used a Web site (www.maff.gov.uk) with an interactive map (fig. 10.4), updated daily, to report the epidemic's extent and provide a detailed description of restrictions. GIS also proved useful in disposing of infected carcasses and assessing damage to farms.

Britain's experience with foot-and-mouth disease demonstrates the importance of readily available spatial data and the need for timely communication among government departments. In a London *Times* article highlighting the importance of GIS in coping with the epidemic, an unnamed GIS specialist criticized the lack of information sharing among government units. "The classic case was

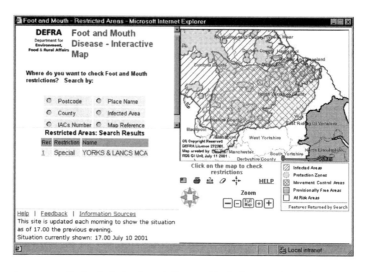

FIGURE 10.4 During the foot-and-mouth disease epidemic of 2001, an interactive map afforded Britons a detailed description of restricted areas. The original color map is from the Department for Environment, Food, and Rural Affairs, "Foot and Mouth Disease— Cases and Infected Areas," http://www.defra.org.

the burying of cattle in a site where it polluted underground wells supplying local houses and farms. That information would have been available but [was] in another department's database." This reluctance of British bureaucrats to share data is hardly unique. Information is not only a source of power but also a potential cause of embarrassment if the data aren't perfect. Adding to bureaucrats' fears is the likelihood of mistakes when databases are incompatible in currentness or level of detail. Haphazard ad hoc data sharing can be worse than no data sharing at all.

+   +   +

Advocates for fuller interagency coordination of geographic data have a new argument: the threat of bioterrorism. The ease of launching a bacterial or viral attack is as worrisome as the potential devastation. Agents range from lethally potent smallpox, which is difficult to obtain but easily cultured, to the temporarily debilitating strain of salmonella with which disciples of Bhagwan Shree Rajneesh contaminated salad bars in The Dalles, Oregon, in 1984. If the goal is disruption—the Rajneeshes wanted to stifle turnout in a local election—a terrorist group need only make its victims sick, not kill them. But if the intent is retribution, woe to anyone who inhales a deadly dose of anthrax spores. That's why the Defense Department vaccinates soldiers for anthrax, which Rear Admiral Michael Cowan, medical readiness director for the Joint Chiefs of Staff, called "the poor man's atom bomb." Also vulnerable is the nation's food supply, which can be deliberately infected with natural agents like foot-and-mouth disease.

Early detection is essential. Unlike bombings and chemical attacks, a bioterrorist assault might not be apparent for days. Smallpox, for instance, generally takes twelve to fourteen days to produce recognizable symptoms and could go unnoticed longer if public health officials lack an efficient network for collecting reports and notifying practitioners. With an incubation period of one to six days, inhalation anthrax is a bit quicker. But the two postal workers who died in the October 2001 anthrax incident did not seek medical attention for more than a week after their apparent exposure at the Brentwood, Maryland, post office. Effective detection includes far more than knowing who's infected and where they live: before health officials can start screening for exposure, administering antibiotics, and isolating likely sources of contamination, they must analyze activity patterns, consumption habits, and other clues to the disease's source and dissemination. Underscoring the need for quick, effective action is an estimate that an anthrax attack would impose a long-term economic burden as large as $26.2 billion for every 100,000 persons exposed.

On the bright side, if there is one, enhanced surveillance will also defend against potent new natural threats like hantavirus, West Nile virus, and antibiotic-resistant bacteria. According to Donald Henderson, director of the Center for Civilian Biodefense Strategies at the Johns Hopkins University, "whether the source is bioterrorism or a naturally occurring outbreak, similar resources are needed." And as Britain's battle with foot-and-mouth disease has shown, geographic data systems are a key part of this defense, particularly in managing quarantine and vaccination programs. Rapid response requires integration of data on the areal extent of exposure, the locations of treatment facilities and supplies of vaccine, and transport nodes at which quarantine is most effectively enforced. Effective response requires conscientious planning for a wide range of contingencies, including spontaneous evacuation by panicked populations. Although staying inside a well-sealed room might be the best strategy, most people's gut reaction is to get the hell out as quickly as possible.

Geographic modeling is a basic element of disaster planning. Applications include spatial models that let emergency planners simulate the spread of disease agents through a water supply system or the dispersion of airborne contaminants from chemical plants, nuclear power stations, and other attractive terrorist targets. In a similar vein, vulnerability models can help homeland security personnel identify water intakes, nuclear reactors, and other sites that threaten large populations and warrant careful surveillance. Equally valuable are models of contagious diseases with which public health officials can evaluate strategies for stockpiling vaccines as well as assess the relative efficacy of vaccination and quarantine as responses to a smallpox attack. And should terrorists attack, geospatial models can help officials forecast the near-term spread of the contaminant or disease. What's more, computer reconstruction of the incident might prevent further attacks by fingering the perpetrator.

Epilogue:
Locational Privacy
as a Basic Right

If George Orwell had not written *1984* to condemn creeping totalitarianism, Big Brother would be a heck of a brand name. A less politically anxious novelist could have told a futuristic tale of a benign state that reveres surveillance technology in much the same way liberal democrats applaud our "social safety net." Instead of "Big Brother is watching you," the mantra would be "Big Brother is watching over us." And instead of trying to manipulate minds, a warm and fuzzy Super Sibling would try to mollify any hazards it could control and warn of those it couldn't. To provide dramatic tension, the writer could add a few unscrupulous control freaks or money-hungry entrepreneurs—much like what we have today. A suitably odious villain might resemble the Acme Rent-a-Car Company, which put GPS tracking systems in its vehicles and fined unsuspecting customers for speeding. If Acme were truly interested in highway safety, it could have installed speed governors.

As a metaphor for intrusive surveillance, Big Brother is almost everywhere—thanks to Orwell, only systems that monitor weather or forest fires seem to sidestep the pejorative. Some of these fears strike me as silly. Especially ludicrous are libertarian essayists who see red-light cameras and E-ZPass as embodiments of Big Brother. Appropriately placed, widely advertised cameras make drivers more attentive, and I'm all for them, especially after an accident several years ago when I hit a driver who ran a stoplight. Equally advantageous are technologies that reduce delays at tollbooths. That said, I understand objections to systems that photograph everybody as well as the concerns voiced when San Diego and Washington, D.C., hired private firms like Lockheed-Martin to write traffic tickets and collect fines. It's always more reassuring when the folks enforcing laws are accountable to local voters rather than distant shareholders.

More akin to Orwell's Big Brother are video cameras in public places. Used to prevent theft and stifle hooliganism, surveillance cameras are the closest most cities get to 1984. Attitudes toward the cameras vary widely, perhaps because of what privacy advocates call creeping acquiescence. We see video monitors at the 7-Eleven, where they thwart thievery. We see them in office buildings, where a single guard can watch multiple entrances and every elevator. Although England adopted street cameras enthusiastically as a defense against terrorism and rising crime, video surveillance seems a bit out of place on American street corners, largely (I suspect) because local governments are not convinced it's all that useful.

Objections vary. Some people dislike being captured on film or videotape under any circumstances, others consider the cameras a poor substitute for beat cops. Civil libertarians quibble with police about the retention period—the ACLU wants the images erased within two days, the cops think a week is essential, and Chandra Levy's mysterious disappearance in May 2001 created an argument for holding the tapes for at least a month. A very different concern is whether anyone is really minding the monitor. Cities that adopt street cameras but don't watch their screens diligently can expect lawsuits from citizens mugged in places presumed safe. Posted disclaimers, the typical talisman against litigation, would only undermine the alleged power of the cameras' panoptic gaze.

Attitudes changed, momentarily at least, after the September 11, 2001, attack on the World Trade Center killed three thousand people and put tens of thousands more out of work. Realization that "it can happen here" triggered debate on trade-offs between liberty and security. One side warned of a slippery slope, the other observed that the Constitution is not a suicide pact. Although news coverage included video pictures of hijackers passing through the security checkpoint at the Portland, Maine, airport, nothing was said about street cameras, which have little to do with keeping terrorists off airplanes if not out of the country.

Facial biometrics shifts the debate. Image analysis software can reduce a person's face to a set of numbers easily compared with corresponding measurements for fugitives, known troublemakers, and anyone else on the police watch-list. If you're in the system, you can be tracked by an extensive network of cameras programmed to hand a suspect off to the next sector's video vigilantes. Privacy advocates eagerly invoke images of Big Brother while technocrats argue that face-recognition electronics is no more intrusive than a well-coordinated team of sharp-eyed cops with good memories. But what happens when the system confuses an innocent pedestrian with a terrorist bomber or deadbeat dad? I don't know about software, but the police have a poor record of admitting and apologizing for their mistakes. Unless refined facial-recognition systems can filter out known look-alikes, repeated misidentifications seem likely. The unintended consequences of flagrant false alarms might include increased public acceptance of masks and disguises in much the same way that red-light surveillance systems created a market for plastic license-plate covers designed to defeat automatic cameras.

A more cost-effective approach is the placement of retinal-scan imaging systems with high-resolution cameras at airports, where the public not only expects scrutiny but applauds the promise of shorter lines and increased security. Despite decades of libertarian warnings, passengers seem eager for a de facto national identity card with encoded biometric data. To verify a traveler's identity, the system would compare the person's retinal scan with the card's biometric profile. No match, no fly. But because most of us would want one, travel ID cards could become as commonplace and indispensable as driver's licenses, and easily linked to databases revealing far more than where we've been and what we look like.

Orwellian scenarios also encompass dataveillance. Integrated analysis of our purchases, memberships, and financial data can tell a lot about who we are and what we're thinking. Although dataveillance typically is used to sell us more of the same or something just a bit different, this form of manipulation is potentially invasive—a clear violation of Supreme Court Justice Louis Brandeis's classic definition of privacy as "the right to be let alone." Although the Gramm-Leach-Bliley Financial Services Modernization Act of 1999 gave consumers and investors limited rights to control their data, its opt-out procedures are arcane, inconsistent, and unable to guarantee the confidentiality most of us crave. More frightening to many is the heavy-handedness of health insurers who extort access to medical records and credit agencies that sell financial histories to clever private investigators. Privacy advocates face stiff resistance from well-heeled lobbyists abetted by lax legislators and complacent consumers. The controversy is certain to escalate if a de facto national identity card gives merchants a more reliable way to track consumers.

In a real war or a sustained confrontation with terrorists, the niceties of rules and regulations are easily overlooked. A classic example is the Census Bureau's insistence that no employee violate anyone's privacy through a deliberate or careless disclosure of personal data. For the most part, the policy works admirably. But a dirty little secret—if one can call a disclosure affecting thousands of innocent people "little"—is the bureau's support of the Western Defense Command's 1942 "evacuation" to internment camps of over 110,000 Japanese Americans living in the Pacific states. Although census officials refused to surrender household questionnaires, they gave the military what writer Erik Larson called "the next best thing": block-by-block counts of residents with Japanese ancestors. Ironically, because Congress broadened the Voting Rights Act in

1975 to cover Asian and Hispanic minorities, the redistricting data currently available to local politicians and anyone else include block-level counts for Asian Americans and other groups.

Although the Census Bureau ensures that its small-area data don't disclose personal information, address-matching and data-integration software help cluster consultants make vaguely reliable inferences about our neighborhoods, buying habits, and voting preferences. Marketers, fundraisers, and politicians thirsting for finer details accumulate data on purchases, investments, donations, and completed questionnaires, and trade information with other organizations. In the absence of restraints on data exchange, data warehouses construct personal profiles that help pitchmen clog our mailboxes and answering machines with unwanted solicitations. A promising solution, which data vendors dread, is an opt-in requirement whereby no one can sell or trade our records without our explicit permission. As an added safeguard, lawmakers should let us control the use of our addresses (both home and e-mail) and Social Security numbers as common links for cross-matching data collected for diverse purposes.

Lawmakers might also set guidelines for locationally precise disease and crime maps. Although data aggregated by census tract or block group generally afford suitable anonymity, maps that pinpoint cancer or crime victims could prove troubling to injured parties and other members of their households. Health advocacy groups and law enforcement agencies that make and circulate maps should be especially wary of map hackers, who can scan a pin map and convert its dot symbols to addresses. When victims want to be left alone, a detailed map could add to their pain.

Privacy rights occasionally conflict with the public access requirements of fair and open government. If you live in an area where property taxes are based on fair market value, anyone can find out how much you paid for your house, what the assessor thinks it's worth, and how many bathrooms you have. If this seems invasive, get over it—otherwise you lose a valuable tool in appealing an unfair or inaccurate assessment. Your only complaint might be with an assessment department that offers Internet access to public records that once required a tedious visit to city hall or the courthouse. However repugnant the thought of curious folks thousands of miles away browsing the local tax rolls, online access can be especially helpful to future neighbors eager to scope out the housing market. Like most forms of geospatial surveillance, online tax rolls are ethically ambiguous. And then there's the view from above. Under the open field doctrine, whatever can be seen with commonly available technology is fair game for official or unofficial snooping. Although thermal imaging of a home's walls or roof without a warrant is now prohibited, overhead imaging of our farms or backyards is fully constitutional. Unless specifically prevented from doing so, governments can use air photos or satellite imagery to troll for illegal crops, unauthorized use of irrigation water, nonconforming land use, and a host of zoning or building-code violations, including unapproved additions, outbuildings, or swimming pools. And in much the same way that countries spy on each other, corporations can monitor competitors' loading docks, parking lots, and outdoor facilities for potentially revealing differences. Overhead spying is legal as well as practicable because an energetic commercial remote sensing industry has made high-resolution color-infrared imagery sufficiently commonplace and affordable to meet judicial standards for "commonly available"—a moving target, to say the least. What bears watching is the possibility that hyperspectral imaging will breach the limits of reasonableness and raise judicial hackles in doing so.

Although remote sensing is a few decades older than global positioning, GPS-based tracking has become far more common. In addition to pinpointing cell-phone users who push 911 as well as anyone merely walking around with a wireless device in roaming mode, satellite tracking could revolutionize criminal justice. All but the least trustworthy sex-cons could enjoy a new start unfettered by shaming, and more vigilant monitoring of parolees could cut recidivism for less abhorrent crimes. Be wary, though, of unintended consequences. Although satellite tracking might reduce prison time for minor offenders—and the likelihood of their picking up more bad habits—some electronic detainees are certain to snap their digital leashes. Equally daunting is the possibility of mainstreaming prisoners poorly prepared to reenter society—recall the mixed results in the 1960s and 1970s, when states eager to slash their mental health budgets closed psychiatric hospitals but skimped on outpatient treatment? Far scarier, a surplus of prison cells could join electronic monitoring as an incentive for expanding the list of felonies and misdemeanors. Imagine what Orwell could have done had he foreseen GPS? A real-time map of dissidents seems far more intriguing than Big Brother's ubiquitous wall screens, especially when engineers figure out a reliable, low-cost way of making GPS work indoors.

A technology to watch is the Digital Angel, which can develop in various ways, some beneficial, others frightening. GPS monitors could no doubt prove useful in tracking Alzheimer's patients, retrieving errant pets, and finding lost children in much the same manner that OnStar helps motorists who break down or crash their Cadillacs. But recall the original patent's reliance on chip implants and sensors for monitoring vital signs: a reasonable enhancement insofar as Grandpa's heart condition is at least as important as his location. Even so, it's a small step technologically if not ethically to letting parents equip their kids with a subcutaneous GPS linked to a GIS that sends a pinch or pain command when Sis or Junior steps (literally) out of bounds. Move over, Big Brother, and make room for Big Momma.

Potentially more contentious is the technological union of GPS, GIS, and everyday wireless communications. The GPS chip in your cell phone calculates your coordinates, which the wireless network forwards to a tracking center, where a GIS puts you on its map so that the police, fire service, or medics can find you in an emergency. In addition to relating a would-be victim's location to the street grid, the GIS can tell parents when a child is in the wrong place, an employer when a worker is not in the right place, and a merchant when a potential customer is nearby. Satellite tracking makes location a commodity, which the "location-based services" (LBS) industry is eager to sell to anyone concerned about where you are or where you've been. And because LBS can be enormously convenient when you need to find something, you're both a part of the product and a potential consumer.

Until recently no one spoke of "locational privacy"—before one's whereabouts was so easily determined, archived, and sold, locational nakedness was hardly an issue. This newness in no way diminishes location's status as a privacy right—all notions of privacy are social responses to innovative technologies that screen or intrude. Screening technologies that radically altered standards of personal privacy include the primitive textiles that made it easy to hide one's genitals and whatever rudimentary walls helped our distant ancestors dress in private and defecate in solitude. More recent examples are the sealed envelope and the front-door peephole. By contrast, intrusive technologies include the telephone and computerized financial records, which led to restrictions, however imperfect, on telemarketing and data exchange. And as call blocking and Caller ID illustrate, intrusive technologies often stimulate screening innovations, which in turn raise questions about cost and permeability. In this typology, wireless location tracking is an intrusive technology conveniently controlled by electronic screening.

Whether locational privacy emerges as a basic human right will depend on the inevitable battle between privacy advocates and industry lobbyists. Privacy advocates have the newness of LBS on their side as well as frightening and wholly plausible scenarios of GPS-based stalking and annoying sales pitches. Although industry lobbyists might claim otherwise, LBS is ripe for opt-in restrictions whereby wireless providers cannot sell or archive our locational data unless we let them and cannot reveal our locational history to public safety officials unless we agree or a judge signs a warrant. Without opt-in restrictions, some users, I'm sure, will assert control by leaving their cell phones at home or by removing the batteries: a strategy that erodes the effectiveness of E-911 and invites fatal consequences. For some of us, Big Business is a worse threat than Big Brother.

Notes

Chapter 1. Maps That Watch

General Sources

Insightful assessments of spatial privacy include Michael R. Curry, "The Digital Individual and the Private Realm," *Annals of the Association of American Geographers* 87 (1997): 681–99; Jerry Dobson, "Is GIS a Privacy Threat?" *GIS World* 11 (July 1998): 34–35; Jon Goss, "'We Know Who You Are and We Know Where You Live': The Instrumental Rationality of Geodemographic Systems," *Economic Geography* 71 (1995): 171–98; and Harlan J. Onsrud, Jeff P. Johnson, and Xavier Lopez, "Protecting Personal Privacy in Using Geographic Information Systems," *Photogrammetric Engineering and Remote Sensing* 60 (1994): 1083–95. Useful overviews of geographic information systems include Nicholas Chrisman, *Exploring Geographic Information Systems,* 2nd ed. (New York: John Wiley and Sons, 2001); Keith C. Clarke, *Getting Started with*

Geographic Information Systems, 3rd ed. (Upper Saddle River, N.J.: Prentice-Hall, 2001); and Paul A. Longley, Michael F. Goodchild, and David J. Maguire, eds., *Geographical Information Systems: Principles, Techniques, Applications, and Management*, 2nd ed., 2 vols. (New York: John Wiley and Sons, 1999). Key references on remote sensing include John R. Jensen, *Remote Sensing of the Environment: An Earth Resource Perspective* (Upper Saddle River, N.J.: Prentice Hall, 2000); and Thomas M. Lillesand and Ralph W. Kiefer, *Remote Sensing and Image Interpretation*, 4th ed. (New York: John Wiley and Sons, 2000). For a history of the Landsat program, see Donald T. Lauer, Stanley A. Morain, and Vincent V. Salomonson, "The Landsat Program: Its Origins, Evolution, and Impacts," *Photogrammetric Engineering and Remote Sensing* 63 (1997): 36–38; Pamela E. Mack, *Viewing the Earth: The Social Construction of the Landsat Satellite System* (Cambridge, Mass.: MIT Press, 1990), esp. 38; and Ray A. Williamson, "The Landsat Legacy: Remote Sensing Policy and the Development of Commercial Remote Sensing," *Photogrammetric Engineering and Remote Sensing* 63 (1997): 877–85.

For general information about GPS, see Michael Ferguson, *GPS Land Navigation: A Complete Guidebook for Backcountry Users of the Navstar Satellite System* (Boise, Id.: Glassford Publishing, 1997); and Gregory T. French, *Understanding the GPS: An Introduction to the Global Positioning System* (Bethesda, Md.: GeoResearch, 1996). For the technical details of GPS, see American Society of Civil Engineers, *Navstar Global Positioning System Surveying* (Reston, Va.: ASCE Press, 2000); Elliot D. Kaplan, ed., *Understanding GPS: Principles and Applications* (Boston: Artech House, 1996); Michael Kennedy, *The Global Positioning System and GIS* (Chelsea, Mich.: Ann Arbor Press, 1996); and National Research Council, Committee on the Future of the Global Positioning System, *The Global Positioning System: A Shared National Asset* (Washington, D.C.: National Academy Press, 1995).

Notes

4   For a concise introduction to dispersion modeling, see Mark Monmonier, *Cartographies of Danger: Mapping Hazards in America* (Chicago: University of Chicago Press, 1997), 127–47, 161–67, 225–29.

4   For a concise overview of address-oriented GIS applications, see William J. Drummond, "Address Matching: GIS Technology for Mapping Human Activity Patterns," *Journal of the American Planning Association* 61 (1995): 240–51.

6   For a history of the TIGER concept, see Donald F. Cooke, "Topology and TIGER: The Census Bureau's Contribution," in *The History of Geographic Information Systems: Perspectives from the Pioneers*, ed. Timothy W. Foresman (Upper Saddle River, N.J.: Prentice Hall, 1998), 47–57.

6   For more information on the use of block data in congressional redistricting, see Mark Monmonier, *Bushmanders and Bullwinkles: How Politicians*

Manipulate Electronic Maps and Census Data to Win Elections (Chicago: University of Chicago Press, 2001), esp. 96–98.

7   For a concise summary of Census Bureau regulations on privacy and nondisclosure, see Michael R. Lavin, *Understanding the Census: A Guide for Marketers, Planners, Grant Writers, and Other Data Users* (Kenmore, N.Y.: Epoch Books, 1996), 10–11, 44–45, 363–65.

7   As its name implies, a block group is an areal unit formed by combining several contiguous blocks. A census tract has a maximum of nine block groups, each of which contains between 250 and 550 housing units and between 1,000 and 1,200 people. For a description of small geographical units used by the Census Bureau, see Lavin, *Understanding the Census,* 150–93.

7   Although fear of rapists . . . : Robert L. Jacobson, "'Megan's Laws' Reinforcing Old Patterns of Anti-Gay Police Harassment," *Georgetown Law Journal* 87 (July 1999): 2431–73, esp. 2456.

7   New York State Sex Offenders Registry, http://criminaljustice.state.ny.us/nsor/index.htm. Before New York State opened its official online registry, a private group, Parents for Megan's Law, set up its own Web site (www.parentsformeganslaw.com) with information copied from the state registry by parent volunteers; see Sue Weibezahl, "Group's Web Site Lists Sex Convicts," *Syracuse Post-Standard,* May 2, 2000.

7   Like Web sites maintained . . . : See "Offenders Online," *Government Computer News State and Local* 6, no. 6 (March 2000): 6; and Paul Zielbauer, "Posting of Sex Offender Registries on Web Sets Off Both Praise and Criticism," *New York Times,* May 22, 2000. New Jersey's efforts to place sex-offender information on the Internet encountered substantial legal resistance; see David Kocieniewski, "Amendment Would Let State Name Sex Offenders Online," *New York Times,* March 28, 2000; and "A Court Blocks Disclosures about Sex Offenders under 'Megan's Law,'" *New York Times,* April 19, 2000.

7   For critiques of publicly accessible sex-offender registries, see Reese Dunklin, David Heath, and Julie Lucas, "About 700 Sex Offenders Do Not Appear to Live at the Addresses Listed on a St. Louis Registry," *St. Louis Post-Dispatch,* May 2, 1999; and Susan R. Paisner, "Exposed: Online Registries of Sex Offenders May Do More Harm Than Good," *Washington Post,* February 21, 1999. For a state-by-state summary of practices, see Alan R. Kabat, "Scarlet Letter Sex Offender Databases and Community Notification: Sacrificing Personal Privacy for a Symbol's Sake," *American Criminal Law Review* 35 (winter 1998): 333–70.

8   Several years ago . . . : See the State Department of Assessments and Taxation, www.dat.state.md.us/sdatweb/.

8   For further information on the Emergency Planning and Community Right-to-Know Act, see Gary D. Bass and Alair MacLean, "Enhancing the

Public's Right-to-Know about Environmental Issues," *Villanova Environmental Law Journal* 4 (1993): 287–321; and Ute J. Dymon, "Mapping: The Missing Link in Reducing Risk under SARA III (Emergency Planning and Community Right-To-Know)," *Risk: Health, Safety, and Environment* 5 (1994): 337–49.

8 The act makes . . . : William J. Craig and Sarah A. Elwood, "How and Why Community Groups Use Maps and Geographic Information," *Cartography and Geographic Information Systems* 25 (1998): 95–104.

10 For reports on the 5-foot resolution of the KH-4B imaging system, used between 1967 and 1972, see Philip J. Klass, "CIA Reveals Details of Early Spy Satellites," *Aviation Week and Space Technology* 142 (June 12, 1995): 167–68; and Kevin C. Ruffner, ed., *Corona: America's First Satellite Program* (Washington, D.C.: Central Intelligence Agency, Center for the Study of Intelligence, 1995), xv. Another authority suggests that the resolution was closer to 2 meters (6 ft.); see Robert A. McDonald, "Corona: Success for Space Reconnaissance, a Look into the Cold War, and a Revolution for Intelligence," *Photogrammetric Engineering and Remote Sensing* 61 (1995): 689–719, esp. 691.

10 For additional information on Système pour l'Observation de la Terre (SPOT), see "SPOT Earth Resources Satellite Beginning Commercial Operation," *Aviation Week and Space Technology* 124 (May 5, 1986): 101; Christopher P. Fotos, "Commercial Remote Sensing Satellites Generate Debate, Foreign Competition," *Aviation Week and Space Technology* 129 (December 19, 1988): 48; and Paul M. Treitz, Philip J. Howarth, and Peng Gong, "Application of Satellite and GIS Technologies for Land-Cover and Land-Use Mapping at the Rural-Urban Fringe: A Case Study," *Photogrammetric Engineering and Remote Sensing* 58 (1992): 439–48.

11 . . . Russia's space agency . . . : Trudy E. Bell, "Remote Sensing," *IEEE Spectrum* 32 (March 1995): 24–31.

11 According to the trade journal . . . : Craig Covault, "USAF Eyes Advanced Russian Military Reconnaissance Imagery," *Aviation Week and Space Technology* 140 (May 23, 1994): 53. For discussion of the government's mixed views on Russian efforts to sell satellite imagery, see James R. Asker, "High-Resolution Imagery Seen as Threat, Opportunity," *Aviation Week and Space Technology* 140 (May 23, 1995): 51–53.

11 America recovered . . . : Vernon Loeb, "Spy Satellite Will Take Photos for Public Sale; Launch Allows Company to Market Images of Almost Anywhere," *Washington Post*, September 25, 1999. Two other U.S. firms, Earthwatch and Orbital Imaging, were eager to share the high-resolution commercial remote sensing market; see Joseph C. Anselmo, "Competitors Chasing Ikonos into Orbit," *Aviation Week and Space Technology* 152 (January 31, 2000): 57. Earthwatch's successor company, DigitalGlobe, successfully launched its QuickBird satellite on October 18, 2001, and an-

nounced the imminent availability of 61-centimeter black-and-white and 2.44-meter color imagery. See "Colorado Company's Satellite Launched from California," *Associated Press State and Local Wire*, October 18, 2001. On April 27, 1999, the attempted launch of Ikonos-1 failed after a "payload fairing" (protection cover) did not separate from the rocket; see "Ikonos 1 Fails to Reach Orbit," *GeoWorld* 12 (June 1999): 12. Pronounced "eye-KOH-nos," *Ikonos* is derived from the Greek word for *image*.

11   Almost all, that is . . . : Dee Ann Divis, "Shutter Control Rattles Industry," *Geo Info Systems* 8 (September 1998): 14–16; Michael R. Gordon, "Pentagon Corners Output of Special Afghan Images," *New York Times*, October 19, 2001; and Robert Wright, "Private Eyes," *New York Times Magazine*, September 5, 1999, 50–55.

11   Intelligence satellites . . . : William E. Burrows, who no doubt relied heavily on CIA sources in writing his pioneering account of Corona, forecasts a pixel of 3 inches; see Burrows, *Deep Black: Space Espionage and National Security* (New York: Random House, 1986), 207. Arms control authority Kosta Tsipis also cites a 3-inch pixel; see Kosta Tsipis, "Arms Control Pacts Can Be Verified," *Discover* 8 (April 1987): 78–93. Journalist Howard Hough, who follows intelligence developments, reports a resolution between 3 and 6 inches; see Howard Hough, *Satellite Surveillance* (Port Townsend, Wash.: Loompanics Unlimited, 1991), esp. 16–19, 78, 180, and 186. Obsessive about secrecy, the NRO was reluctant to admit their systems' existence, much less comment on their spatial resolution; see Jeffrey Richelson, *America's Secret Eyes in Space: The U.S. Keyhole Spy Satellite Program* (New York: Harper and Row, 1990), 257–71. More recently, the Federation of American Scientists, at its Space Policy Project Web site, accords the KH-12 system, known as Improved Crystal and described as a downward-looking version of the Hubble Space Telescope, a resolution of 10 centimeters—about 3 inches; see FAS, "Improved Crystal," http://www.fas.org/spp/military/program/imint/kh-12.htm.

11   And airplanes as well . . . : Craig Covault, "Radar Flight Meets Mapping Goals," *Aviation Week and Space Technology* 152 (February 28, 2000): 43.

13   Just ask the former clients . . . : Joe Sharkey, "Most Car Rental Customers Can Relax: The Top Companies Have No Plans to Monitor Speeders," *New York Times*, July 11, 2001.

14   The twenty-four-satellite configuration in figure 1.8 does not show replacement satellites. Because satellites fail or become obsolete, the constellation also includes one or two spares, which the Air Force can reposition as required. With no air resistance in space, tiny but precisely controlled thruster rockets let the Air Force change a satellite's orbit as well as fine-tune its position within an orbit. For an example of thruster technology, see Michael A. Dornheim, "Xenon Thruster to Propel New Millennium Spacecraft," *Aviation Week and Space Technology* 150 (Septem-

ber 25, 1995): 110. Because worldwide GPS will work with fewer than twenty-four satellites, three members of the twenty-four-satellite constellation are sometimes called "active on-orbit spares." For an example, see *Navstar GPS User Equipment Introduction*, public release version (Alexandria, Va.: U.S. Coast Guard Navigation Center, September 1996), 5.

14 Because taxpayers paid . . . : Scott Pace and others, *The Global Positioning System: Assessing National Policies* (Santa Monica, Cal.: Rand Corporation, Critical Technologies Institute, 1995).

14 For additional discussion of GPS error, see *Navstar GPS User Equipment Introduction*, 16–21. According to James R. Clynch of the Naval Postgraduate School Department of Oceanography, PPS and SPS accuracy standards are not fully comparable; see Clynch, "GPS Accuracy Levels," http://www.oc.nps.navy.mil/~jclynch/gpsacc.html.

15 What's more, civilian users . . . : Even with SA, some civilian users could reduce positioning error to 1 meter through a technique called differential GPS. Users needed to link electronically to a nearby base station, the location of which having been established by a highly precise independent survey. Differences between these known coordinates and those estimated by a GPS receiver at the same spot could then be used to adjust readings for other GPS receivers in the vicinity. For a concise explanation, see Thomas A. Herring, "The Global Positioning System," *Scientific American* 274 (February 1996): 44–50; and Matt Lake, "Pentagon Lets Civilians Use the Best G.P.S. Data," *New York Times*, June 15, 2000.

15 "with SA activated . . . ": National Geodetic Survey, "Comparison of Positions with and without Selective Availability," http://www.ngs.noaa.gov/FGCS/info/sans_SA/compare/DSRC.htm. With SA turned off, differential GPS provides even more accurate coordinates.

15 Even so, the Pentagon . . . : Dee Ann Divis, "SA No More: GPS Accuracy Increases 10 Fold," *Geospatial Solutions* 10 (June 2000): 18–20; and Carla Anne Robbins, "Government Will Authorize Civilian Use of Military-Quality Positioning Signal," *Wall Street Journal*, May 1, 2000.

15 "selectively deny GPS . . . " : "Clinton Statement on Global Positioning System Accuracy," U.S. Newswire, May 1, 2000. The military may have invoked regional denial during the 2001 Afghanistan campaign; see Declan McCullagh, "U.S. Could Deny GPS to Taliban," *Wired News*, October 20, 2001, http://www.wired.com/news/conflict/0,2100,47739,00.html.

15 Although GPS surveillance . . . : Jerry Dobson, "What Are the Ethical Limits of GIS?" *GEO World* 13 (May 2000): 24–25; and Richard Stenger, "Tiny Human-Borne Monitoring Device Sparks Privacy Fears," *CNN.com*, December 20, 1999, http://www.cnn.com/1999/TECH/ptech/12/20/implant.device/.

Chapter 2. Overhead Assets

General Sources

Useful references for the early history of aerial reconnaissance and mapping include Teodor J. Blachut and Rudolf Burkhardt, *Historical Development of Photogrammetric Methods and Instruments* (Falls Church, Va.: American Society of Photogrammetry and Remote Sensing, 1989); Grover Heiman, *Aerial Photography: The Story of Aerial Mapping and Reconnaissance* (New York: Macmillan, 1972); and Harold E. Porter, *Aerial Observation: The Airplane Observer, the Balloon Observer, and the Army Corps Pilot* (New York: Harper and Brothers, 1921). For satellite remote sensing, my key source was Thomas M. Lillesand and Ralph W. Kiefer, *Remote Sensing and Image Interpretation*, 4th ed. (New York: John Wiley and Sons, 2000).

Principal references for the Corona intelligence satellite program are William E. Burrows, *Deep Black: Space Espionage and National Security* (New York: Random House, 1986); John Cloud and Keith Clarke, "To Do the Other Things: Corona and the Secret Cartography of the Cold War," *ACSM Bulletin* no. 191 (May/June 2001): 25–30; Dwayne A. Day, John M. Logsdon, and Brian Latell, eds., *Eye in the Sky: The Story of the Corona Spy Satellites* (Washington, D.C.: Smithsonian Institution Press, 1998); Kenneth E. Greer, "Corona," *Studies in Intelligence* 17, supplement (spring 1973): 1–37; Philip J. Klass, "CIA Reveals Details of Early Spy Satellites," *Aviation Week and Space Technology* 142 (June 12, 1995): 167–73; Philip Klass, *Secret Sentries in Space* (New York: Random House, 1971); Robert A. McDonald, "Corona: Success for Space Reconnaissance, a Look into the Cold War, and a Revolution for Intelligence," *Photogrammetric Engineering and Remote Sensing* 61 (1995): 689–719; Jeffrey T. Richelson, *America's Secret Eyes in Space: The U.S. Keyhole Spy Satellite Program* (New York: Harper and Row, 1990); Jeffrey Richelson, "The Keyhole Satellite Program," *Journal of Strategic Studies* 7 (1984): 121–53; Kevin C. Ruffner, ed., *Corona: America's First Satellite Program* (Washington, D.C.: Central Intelligence Agency, Center for the Study of Intelligence, 1995); and Seth Shulman, "Code Name: Corona," *Technology Review* 99 (October 1996): 22–32.

The Federation of American Scientists Web site (www.fas.org) was a rich source of information about Corona, its various successors, and current satellite intelligence. John Pike, who created content for and maintained the FAS Web site until late 2000, now pursues his interest in satellite intelligence as director of GlobalSecurity.org. For insights on contemporary satellite intelligence, I also relied on Jeffrey T. Richelson, *A Century of Spies: Intelligence in the Twentieth Century* (New York: Oxford University Press, 1995); and Jeffrey T. Richelson, *America's Space Sentinels: DSP Satellites and National Security* (Lawrence, Kans.: University of Kansas Press, 1999).

Notes

17 As a synonym for aerial or space reconnaissance, "overhead assets" is well established in the military and intelligence communities. For instance, in a March 1996 NATO press briefing on Implementation Force (IFOR) activities in the former Yugoslavia, an officer identified only as Colonel Bryan used the term twice in one paragraph in referring to satellite systems operated by the U.S. National Reconnaissance Office: "Overhead assets are available to us as I have said. IFOR does not directly control overhead assets as much as those products that come from national sources that are fed into the various forces that are under IFOR." IFOR Air Reconnaissance Brief, March 9, 1996, NATO, http://www.nato.int/ifor/trans/t960309b.htm.

18 The French army . . . : Heiman, *Aerial Photography*, 5–9.

18 The Union army . . . : Ibid., 17.

18 As long as a pilot . . . : Improved guns and shells increased the reach of antiaircraft fire from around 6,000 feet early in World War I to over 18,000 feet near the war's end. See Porter, *Aerial Observation*, 111–12, 158–62.

18 For detailed discussion of three-lens cameras used for reconnaissance mapping, see U.S. Army Air Force, "Reconnaissance Mapping with Trimetrogon Photography," in *Manual of Photogrammetry*, preliminary edition, ed. P. G. McCurdy and others (New York: Pitman Publishing Co.; Washington, D.C.: American Society of Photogrammetry, 1944), 645–712. Used heavily during World War II for photointerpretation and reconnaissance mapping, multilens cameras predated the use of aircraft; see Blachut and Burkhardt, *Historical Development of Photogrammetric Methods*, 38–40. Single-lens panoramic cameras, with a wide sweep to the left and right, were also used for photointerpretation; see Francis H. Moffitt and Edward M. Mikhail, *Photogrammetry*, 3rd ed. (New York: Harper and Row, 1980), 75–79.

19 During the mid-1940s . . . : Sidney A. Tischler, "Procedural Developments in Trimetrogon Compilation," *Photogrammetric Engineering* 14 (1948): 53–60; and "Report of Commission I—Photography, to the Sixth International Photogrammetry Congress and Exhibition," *Photogrammetric Engineering* 14 (1948): 229–79, esp. 229–30. To promote useful comparison, I have converted most post-1960 terrain measurements to metric units. For authenticity and mildly harmless inertia, earlier altitudes and areas remain in feet and square miles. Camera measurements retain the customary units of millimeters for film size and inches for focal length.

19 Commercial satellites like SPOT . . . : Because their orbits are inclined away from the poles, low-altitude satellites like Landsat, SPOT, and Ikonos typically ignore or provide less accurate coverage of polar areas.

For an overview of orbital geometry in remote sensing, see Lillesand and Kiefer, *Remote Sensing and Image Interpretation*, 379–83.

19   By contrast, Ikonos offers . . . : Flexible scanners offer different levels of resolution with somewhat different return frequencies. With a very narrow ground swath, Ikonos has a revisit time so large—and so irrelevant—that its owner, Space Imaging, does not publish a hypothetical no-tilt revisit time. For additional details on orbits, resolution, and revisit times, see the technical specifications pages at the Space Imaging Web site, http://www.spaceimaging.com.

21   During the 1930s . . . : Blachut and Burkhardt, *Historical Development of Photogrammetric Methods*, 49–136.

21   Imaging software . . . : See N. Al-Rousan and others, "Automated DEM Extraction and Orthoimage Generation from SPOT Level 1B Imagery," *Photogrammetric Engineering and Remote Sensing* 63 (1997): 965–74; and Rongxing Li, "Potential of High-Resolution Satellite Imagery for National Mapping Products," *Photogrammetric Engineering and Remote Sensing* 64 (1998): 1165–70. For discussion of general principles of orthophotography, see Lillesand and Kiefer, *Remote Sensing and Image Interpretation*, 170–74; and Kurt Novak, "Rectification of Digital Imagery," *Photogrammetric Engineering and Remote Sensing* 58 (1992): 339–44. For an early essay on the principles of satellite-enabled stereovision, see Roy Welch and Wayne Marko, "Cartographic Potential of a Spacecraft Line-Array Camera System: Stereosat," *Photogrammetric Engineering and Remote Sensing* 47 (1981): 1173–85.

22   On August 18 . . . : Klass, "CIA Reveals Details."

23   *Discoverer* was a clever . . . : According to William Burrows, although press releases stressed scientific experiments, only one Discoverer mission actually carried live animals. See Burrows, "Imaging Space Reconnaissance Operations during the Cold War: Cause, Effect, and Legacy," *Cold War Forum*, http://webster.hibo.no/asf/Cold_War/report1/williame.html.

24   The orbit was not . . . : Data are chiefly from McDonald, "Corona: Success for Space Reconnaissance," 691–95; and FAS, "KH-1 Corona," http://www.fas.org/spp/military/program/imint/kh-1.htm. Discoverer-14's elliptical orbit, similar to other Corona missions, varied from 803 kilometers (434 nautical miles) to 177 kilometers (96 naut. mi.). See *Encyclopedia Astronautica*, s.v. "KH-1," http://www.astronautix.com/craft/kh1.htm. The KH-1 camera used in early Corona missions was the optical equivalent of an electronic imaging system with a resolution of 40 feet.

24   As the satellite moved . . . : Sources do not report a ground swath, which would vary somewhat because of the elliptical orbit. For a panoramic scan of 70 degrees and a low altitude of 180 kilometers, trigonometry suggests a minimum swath of 252 kilometers.

24   The first Corona . . . : Because aircraft reconnaissance, still used outside
     the Soviet Union, had been given the code name Talent, more general ref-
     erences to top-secret photography typically adopted the combined name
     Talent-Keyhole.

25   Two stellar cameras . . . : For information on the camera system, see *The
     KH-4B Camera System*, available on the Federation of American Scientists
     Web site, http://www.fas.org/spp/military/program/imint/kh-4%20
     camera%20system.htm. (Now declassified, this document was originally
     a National Reconnaissance Office Data Book [Washington, D.C.: National
     Photographic Interpretation Center, September 1967].) The KH-3, KH-4,
     and KH-4A imaging systems also included one or more supplementary
     cameras. Corona used two other imaging systems: between 1961 and
     1964, twelve missions used a KH-5 camera, designed for small-scale geo-
     detic positioning, and three missions carried a KH-6 camera, intended for
     highly detailed intelligence monitoring. CIA records describe six of the
     twelve KH-5 missions as successful, and one of the KH-6 missions as partly
     successful. Intended to provide 2-foot resolution, the KH-6 system never
     surpassed the KH-4B and was discontinued. For additional information,
     see NASA Jet Propulsion Laboratory Mission and Spacecraft Library,
     "Corona," http://samadhi.jpl.nasa.gov/msl/Programs/corona.html.

25   "scientifically or environmentally useful . . . ": William J. Clinton, "Re-
     lease of Imagery Acquired by Space-Based National Reconnaissance Sys-
     tems (Executive Order 12951, of February 22, 1995)," *Federal Register* 60
     (1995): 10789–90. For an example of a scientific study based on declas-
     sified Corona imagery, see Robert Bindschadler and Patricia Vornberger,
     "Changes in the West Antarctic Ice Sheet since 1963 from Declassified
     Satellite Photography," *Science* 279 (1998): 689–92.

25   For discussion of Corona's role during the Cuban missile crisis, see Bur-
     rows, *Deep Black*, 132–37.

26   Corona not only . . . : Dwayne A. Day, "Mapping the Dark Side of the
     World—Part 1: The KH-5 Argon Geodetic Satellite," *Spaceflight* 40 (1998):
     264–69; Day, "Mapping the Dark Side of the World—Part 2: Secret Geo-
     detic Programmes after Argon," *Spaceflight* 40 (1998): 303–10; and
     Muneendra Kumar, "World Geodetic System 1984: A Reference Frame
     for Global Mapping, Charting, and Geodetic Applications," *Surveying and
     Land Information Systems* 53 (1993): 53–56.

26   Less well known . . . : Day, Logsdon, and Latell, *Eye in the Sky*, 211–14; and
     Richelson, *America's Secret Eyes in Space*, 268–69.

27   "most cherished hope": Quoted in Don Oberdorfer, "SALT: The Tortuous
     Path," *Washington Post*, May 11, 1979.

27   "photoreconnaissance satellites . . . ": Quoted in Edward Walsh, "Carter
     Vows U.S. Will Continue Leadership in Space," *Washington Post*, October
     2, 1978. Also see Richelson, "Keyhole Satellite Program," esp. 121.

27 Well-placed . . . informants . . . : Richelson describes one ostensibly inad-
vertent leak of satellite photos in the transcript of a congressional hearing
on arms control. See Jeffrey T. Richelson, "The Future of Space Reconnaissance," *Scientific American* 264 (January 1991): 38–44, esp. 44.

27 Espionage trials . . . : For information on Kampiles, see James Ott, "Espionage Trial Highlights CIA Problems," *Aviation Week and Space Technology* 109 (November 27, 1978): 21–23; and Richelson, "Keyhole Satellite
Program," 138–40. For reports on Boyce and Lee, see Robert Lindsey, "Alleged Soviet Spy Testifies He Was Blackmailed after Telling a Friend of
C.I.A. 'Deception' of Australia," *New York Times*, April 27, 1977; Lindsey,
"To Be Rich, Young—and a Spy," *New York Times Magazine*, May 22, 1977,
18–28, 89–94, 106–8; Ott, "Espionage Trial"; and Richelson, "Keyhole
Satellite Program," 138–40.

28 Morison had sent . . . : "Satellite Pictures Show Soviet CVN Towering
above Nikolaiev Shipyard," *Jane's Defence Weekly* 2 (August 11, 1984): 171–
73. Also see Burrows, *Deep Black*, 329–30; George Lardner, Jr., "3 Secret
Photos Called Nothing New to Soviets," *Washington Post*, October 11, 1985;
Warren Richey, "Overreaction to Spy Cases Could Harm U.S. As Much As
Lost Secrets," *Christian Science Monitor*, December 5, 1985; and "Spies
Who Aren't," *Los Angeles Times*, May 22, 1986.

28 "just didn't really tell me . . . ": Quoted in Lardner, "3 Secret Photos."

29 One of Richelson's sources . . . : Klass, *Secret Sentries in Space*, 171–72.

29 "develop a new generation . . . ": "Industry Observer," *Aviation Week and
Space Technology* 96 (February 7, 1972): 9.

29 . . . charged coupled devices (CCDs): John A. Adam, "Counting the Weapons," *IEEE Spectrum* 23 (July 1986): 46–56, esp. 47–48.

30 "greater than the KH-9 . . . ": Richelson, "Keyhole Satellite Program," 137.

30 For an overview of the KH-12 satellite, also known by the code name Im-
proved Crystal, see FAS, "Improved Crystal," http://www.fas.org/spp/
military/program/imint/kh-12.htm. For information on the space tele-
scope imaging system, see Tim Beardsley, "Hubble's Legacy: The Space
Telescope Launches a New Era in Astronomy," *Scientific American* 262
(June 1990): 18–22; and Carolyn Collins Petersen and John C. Brandt,
Hubble Vision: Astronomy with the Hubble Space Telescope (Cambridge:
Cambridge University Press, 1995), esp. 23–27.

30 "space telescope with a rocket": FAS, "Improved Crystal."

30 "an unclassified version . . . ": Quoted in Adam, "Counting the Weapons,"
49.

30 For an excellent discussion of image resolution and other details of satellite reconnaissance, see FAS, "IMINT 101: Introduction to Imagery Intelligence," http://www.fas.org/irp/imint/imint_101.htm.

30 A calculation . . . : Adam, "Counting the Weapons," 49.

30 Although 4-inch resolution . . . : See, for example, D. E. Richardson, "Spy

Satellites: Someone Could Be Watching You," *Electronics and Power* 24 (1978): 573–76.

30 In fact, Pike . . . : See FAS, "Resolution Comparison: Reading Headlines and License Plates," http://www.fas.org/irp/imint/resolve5.htm.

31 Figure 2.6: Image array from FAS, "Resolution Comparison," http://www.fas.org/irp/imint/resolve4.htm.

31 For a discussion of the underlying rationale of image interpretability rating scales, see John M. Irvine and Jon C. Leachtenauer, "A Methodology for Developing Image Interpretability Rating Scales," *Technical Papers of the American Society of Photogrammetry and Remote Sensing Annual Convention,* April 1996, vol. 1, pp. 273–81; and Jon C. Leachtenauer, "National Imagery Interpretability Rating Scales: Overview and Product Description," *Technical Papers of the American Society of Photogrammetry and Remote Sensing Annual Convention,* April 1996, vol. 1, pp. 262–72.

31 Focused on usefulness . . . : Quotations from the rating levels and lists are from FAS, "National Image Interpretability Rating Scales," http://www.fas.org/irp/imint/niirs.htm.

32 "if a picture . . . ": John Pike, "Introducing the FAS Public Eye Initiative" (paper presented at Through the Keyhole: Public Policy Applications of Declassified Corona Satellite Imagery, Federation of American Scientists conference, February 16, 1999), http://www.fas.org/eye/conf9902/trans-pike.htm.

32 Look for even bigger . . . : "0.5-Meter Resolution Approved for Ikonos," *Aviation Week and Space Technology* 153 (November 27, 2000): 49.

32 For a concise explanation of SAR, see Charles Elachi, "Radar Images of the Earth from Space," *Scientific American* 247 (December 1982): 54–61; and Tony Freeman, "What Is Imaging Radar?" NASA Jet Propulsion Laboratory, http://southport.jpl.nasa.gov/desc/imagingradarv3.html. For further details, see Donald R. Wehner, *High Resolution Radar* (Norwood, Mass.: Artech House, 1987).

32 Although estimates of ground resolution . . . : The Federation of American Scientists suggests that the resolution, which might "in principle" be better than 1 meter, is rarely so in practice because of the trade-off between ground resolution and geographic scope; see FAS, "Lacrosse/Onyx," http://www.fas.org/spp/military/program/imint/lacrosse.htm. This assessment is consistent with the estimate of "between 3 and 10 feet" reported in Richelson, "Future of Space Reconnaissance," 39.

33 For all-weather coverage . . . : My principal sources, which do not always agree, include FAS, "Lacrosse/Onyx"; Daniel Charles, "Spy Satellites: Entering a New Era," *Science* 243 (1989): 1541–43; Craig Covault, "NRO Radar, Sigint Launches Readied," *Aviation Week and Space Technology* 149 (September 1, 1997): 22–24; Paul Mann, "Congress Backs Raids, Faults

Strategy," *Aviation Week and Space Technology* 149 (December 21/28, 1998): 124–25; and Richelson, "Future of Space Reconnaissance."

33 According to John Pike . . . : For a concise overview, see Mann, "Congress Backs Raids;" and Mark Thompson, "The Pentagon's Plan," *Time* 153 (April 12, 1999): 48–49.

34 Although each . . . spy satellite . . . : See, for example, John Morrocco, "Gulf War Boosts Prospects for High-Technology Weapons," *Aviation Week and Space Technology* 134 (March 18, 1991): 45–47.

34 According to the *Los Angeles Times* . . . : James Risen and Ralph Vartabedian, "U.S. Launches Costly Overhaul of Spy Satellites," *Los Angeles Times*, September 28, 1995. Also see FAS, "8X Enhanced Imaging System," http://www.fas.org/spp/military/program/imint/8x.htm. This page includes links to assessments of 8X capability.

34 Because large image intelligence satellites can be seen by the naked eye, astronomers can easily track their orbits, as the Federation American Scientists points out with "Where are they now?" links to Heavens-Above .com and other Web sites; see John Pike, FAS, "Space Surveillance," http://www.fas.org/spp/military/program/track/. Also see Vernon Loeb, "Hobbyists Track Down Spies in the Sky," *Washington Post*, February 20, 1999.

34 Early sources of information on nonimaging intelligence satellites include John A. Adam, "Peacekeeping by Technical Means," *IEEE Spectrum* 23 (July 1986): 42–45; the Federation of American Scientists' Military Space Programs Web pages, http://www.fas.org/spp/military/program/index.html; Richelson, *America's Space Sentinels*; and Glenn Zorpette, "Monitoring the Tests," *IEEE Spectrum* 23 (July 1986): 57–66.

35 For additional information on the role of seismic networks in test-ban verification, see Paul G. Richards, "Building the Global Seismographic Network for Nuclear Test Ban Monitoring," *EARTHmatters* [Columbia Earth Institute], fall 1999, 37–40; Paul G. Richards and Won-Young Kim, "Testing the Nuclear Test-Ban Treaty," *Nature* 389 (1997): 782–83; and Kosta Tsipis, "Arms Control Pacts Can Be Verified," *Discover* 8 (April 1987): 78–93.

36 Instead of admitting . . . : Richards and Kim, "Testing the Nuclear Test-Ban Treaty;" R. Jeffrey Smith, "U.S. Officials Acted Hastily in Nuclear Test Accusation," *Washington Post*, October 20, 1997; and Lynn R. Sykes, "Small Earthquake Near Russian Test Site Leads to U.S. Charges of Cheating on Comprehensive Nuclear Test Ban Treaty," F.A.S. Public Interest Report, *Journal of the Federation of American Scientists* 50 (November/December 1997): 1–12.

36 Shortly after the incident . . . : Motivated by a firm belief in technology, my informant was not alone in questioning official claims that the attack was

an accident. In October 1999, for instance, newspapers in London and Copenhagen reported that Chinese military officers claimed the embassy had served Yugoslavian forces as a communications center, which NATO was determined to silence. See Joel Bleifuss, "A Tragic Mistake?" *In These Times* 24 (December 12, 1999): 2–3.

36 Turns out . . . : Nearly a year after the May 7, 1999, bombing, the CIA accepted responsibility. An investigation blamed the target's misidentification on a combination of haste, zeal, ignorant inference, and inadequate maps. See Steven Lee Myers, "C.I.A. Fires Officer Blamed in Bombing of China Embassy," *New York Times*, April 9, 2000; and Myers, "Chinese Embassy Bombing: A Wide Net of Blame," *New York Times*, April 17, 2000.

37 For a nontechnical introduction to antisatellite weapons, see Daniel G. Dupont, "Laser Show," *Scientific American* 278 (January 1998): 44; Dupont, "The Real Star Wars," *Scientific American* 280 (June 1999): 36; and Cynthia A. S. McKinley, "When the Enemy Has Our Eyes" (master's thesis, School of Advanced Airpower Studies, Air University, Maxwell Air Force Base, Ala., 1996).

37 For discussion of international conflict and opportunities arising from high-resolution commercial imagery, see J. Todd Black, "Commercial Satellites: Future Threats of Allies," *Naval War College Review* 52 (winter 1999): 99–114; Vipin Gupta, "New Satellite Images for Sale," *International Security* 20 (summer 1995): 94–125; and Michael J. Riezenman, "Spying for Dummies," *IEEE Spectrum* 36 (November 1999): 62–69.

37 For insights on "shutter control" and federal remote sensing policy, see Joseph C. Anselmo, "Shutter Controls: How Far Will Uncle Sam Go?" *Aviation Week and Space Technology* 152 (January 31, 2000): 55–56; Dee Ann Divis, "Remote Regs, SRTM, and Financing NSDI," *Geo Info Systems* 10 (May 2000): 18–20; House Committee on Science, Space, and Technology and the Permanent Select Committee on Intelligence, *Commercial Remote Sensing in the Post–Cold War Era*, 103rd Cong., 2nd sess., February 9, 1994; and Ray A. Williamson, "The Landsat Legacy: Remote Sensing Policy and the Development of Commercial Remote Sensing," *Photogrammetry and Remote Sensing* 63 (1997): 877–85.

37 "every bad guy . . . ": Quoted in Dan Charles, "Every Move You Make," *New Scientist* 155 (August 2, 1997): 18–19, quotation on 19.

Chapter 3. Eyes on the Farm

General Sources

Key general references on the Agricultural Adjustment Administration (AAA) include Gladys Baker, *Century of Service: The First 100 Years of the United States De-*

partment of Agriculture (Washington, D.C.: Government Printing Office, 1963), 143–78; Donald C. Blaisdell, *Government and Agriculture: The Growth of Federal Farm Aid* (New York: Farrar and Rinehart, 1940), 39–75; Edwin G. Nourse, Joseph S. David, and John D. Black, *Three Years of the Agricultural Adjustment Administration* (Washington, D.C.: Brookings Institution, 1937), esp. 60–77; and *Report of the Secretary of Agriculture, 1936* (Washington, D.C., 1936), 8–17. For information of the AAA's use of aerial photography, I relied on W. N. Brown, "Area Measurements by Use of Aerial Photography," *Photogrammetric Engineering* 2 (January-February-March 1936): 19–22; Ralph H. Moyer, "Use of Aerial Photographs in Connection with Farm Programs Administered by the Production and Marketing Administration, U.S.D.A.," *Photogrammetric Engineering* 15 (1949): 536–40; and Harry Tubis, "Aerial Photography Maps Our Farmlands: The Program of the Agricultural Adjustment Administration," *Photogrammetric Engineering* 3 (April-May-June 1937): 21–23.

General sources for my discussion of soils mapping are Mark Baldwin, Howard M. Smith, and Howard W. Whitlock, "The Use of Aerial Photographs in Soil Mapping," *Photogrammetric Engineering* 13 (1947): 532–36; Edward W. Magruder, "Aerial Photographs and the Soil Conservation Service," *Photogrammetric Engineering* 15 (1949): 517–36; Ralph J. McCracken and Douglas Helms, "Soil Surveys and Maps," in *The Literature of Soil Science,* ed. Peter McDonald (Ithaca, N.Y.: Cornell University Press, 1994), 275–311, esp. 301–4; Roy W. Simonson, "Use of Aerial Photographs in Soil Surveys," *Photogrammetric Engineering* 16 (1950): 308–15; and U.S. Department of Agriculture, Soil Conservation Service, *Aerial-Photo Interpretation in Classifying and Mapping Soils,* Agriculture Handbook no. 294 (Washington, D.C., 1966).

My principal general reference on remote sensing is Thomas M. Lillesand and Ralph W. Kiefer, *Remote Sensing and Image Interpretation,* 4th ed. (New York: John Wiley and Sons, 2000), esp. 89–93 for a discussion of spectral signatures. Useful references focused on remote sensing of soils include Maxwell B. Blanchard, Ronald Greeley, and Robert Goettelman, "Use of Visible, Near-Infrared, and Thermal Infrared Remote Sensing to Study Soil Moisture," *Proceedings of the International Symposium on Remote Sensing of Environment* 9 (1974): 693–700; L. F. Curtis, "Remote Sensing of Soil Moisture: User Requirements and Present Prospects," in *Remote Sensing of the Terrestrial Environment,* ed. R. F. Peel, L. F. Curtis, and E. C. Barrett (London: Butterworths, 1977), 143–58; Lillesand and Kiefer, *Remote Sensing and Image Interpretation,* 230–36; T. Schmugge, "Remote Sensing of Surface Soil Moisture," *Journal of Applied Meteorology* 17 (1978): 1549–57; and John R. Schott, *Remote Sensing: The Image Chain Approach* (New York: Oxford University Press, 1997), 196–211.

General references on remote sensing applications in agricultural include John E. Anderson, Robert L. Fischer, and Stephen R. Deloach, "Remote Sensing and Precision Agriculture: Ready for Harvest or Still Maturing?" *Photogrammetric Engineering and Remote Sensing* 65 (1999): 1118–23; M. S. Moran, "Landsat

TM and ETM+ Data for Resource Monitoring and Management," Basic Science and Remote Sensing Initiative, Michigan State University, http://www.bsrsi.msu.edu/~qi/landsat.html; M. S. Moran and others, "Combining Multifrequency Microwave and Optical Data for Crop Management," *Remote Sensing of Environment* 61 (1997): 96–109; M. S. Moran, Y. Inoue, and E. M. Barnes, "Opportunities and Limitations for Image-Based Remote Sensing in Precision Crop Management," *Remote Sensing of Environment* 61 (1997): 319–46; and Gail Wade, Rick Muehller, and Paul Cook, "AVHRR Map Products for Crop Condition Assessment: A Geographic Information Systems Approach," *Photogrammetric Engineering and Remote Sensing* 60 (1994): 1145–50.

General sources on precision agriculture include R. W. Gunderson and others, "The Collective: GIS and the Computer-Controlled Farm," *Geospatial Solutions* 10 (October 2000): 30–34; National Research Council, Board on Agriculture, Committee on Assessing Crop Yield: Site-Specific Farming, Information Systems, and Research Opportunities, *Precision Agriculture in the 21st Century: Geospatial and Information Technologies in Crop Management* (Washington, D.C.: National Academy Press, 1997); "Precision Agriculture: Information Technology for Improved Resource Use," *Agricultural Outlook* no. 250 (April 1998): 19–23; and J. K. Schueller, "Technology for Precision Agriculture," in *Precision Agriculture 1997*, ed. John V. Stafford (Oxford: BIOS Scientific Publishers, 1997), 33–44.

For information on agricultural applications of GPS, I relied on Thomas Bell, "Automatic Tractor Guidance Using Carrier-Phase Differential GPS," *Computers and Electronics in Agriculture* 25 (2000): 53–66; Wolfgang Lechner and Stefan Baumann, "Global Navigation Satellite Systems," *Computers and Electronics in Agriculture* 25 (2000): 67–85; and J. N. Wilson, "Guidance of Agricultural Vehicles: A Historical Perspective," *Computers and Electronics in Agriculture* 25 (2000): 3–9.

Sources concerned with the social impacts of precision agriculture include Don E. Albrecht, "The Correlates of Farm Concentration in American Agriculture," *Rural Sociology* 57 (1992): 512–20; Rick Welsh, "Vertical Coordination, Producer Response, and the Locus of Control over Agricultural Production Decisions," *Rural Sociology* 62 (1997): 491–507; Dan Whipple, "Seeds of Controversy," *Futurist* 33 (October 1998): 10–12; and Mary Zey-Ferrell and William Alex McIntosh, "Agricultural Lending Policies of Commercial Banks: Consequences of Bank Dominance and Dependency," *Rural Sociology* 52 (1987): 187–207.

Useful sources on the use of remote sensing to detect marijuana grows include C. S. T. Daughtry and C. L. Walthall, "Spectral Discrimination of *Cannabis sativa* L. Leaves and Canopies," *Remote Sensing of Environment* 64 (1998): 192–201; and appendix C of "2000 Counterdrug Research and Development Blueprint Update," Office of National Drug Control Policy, http://www.whitehousedrugpolicy.gov/publications/scimed/blueprint00/appendixc.html.

For insights on constitutional and personal privacy issues raised by aerial imaging, see David Reed, "Thermal Surveillance: Poised at the Intersection of Technology and the Fourth Amendment," Computers and Law, University of Buffalo School of Law, http://wings.buffalo.edu/law/Complaw/ComplawPapers/t.html; E. Terrence Slonecker, Denice M. Shaw, and Thomas M. Lillesand, "Emerging Legal and Ethical Issues in Advanced Remote Sensing Technology," *Photogrammetric Engineering and Remote Sensing* 64 (1998): 589–95; and Lisa J. Steele, "The View from on High: Satellite Remote Sensing Technology and the Fourth Amendment," *Berkeley Technology Law Journal* 6 (1991): 317–34.

Notes

39 "ever-normal granary": Baker, *Century of Service*, 158. In addition to price stabilization and conservation, New Deal farm programs included loans to farmers, drought relief, emergency marketing quotas, free food distribution to the poor, and programs to sell American farm products overseas.

39 "economic democracy": Baker, *Century of Service*, 159.

40 "Before we can . . . ": H. R. Tolley, "Aerial Photography and Agricultural Conservation," transcript, radio interview broadcast December 21, 1937, Department of Agriculture main library.

40 A planimeter is a mechanical device for measuring the area within an irregular closed curve; see D. H. Maling, *Measurements from Maps: Principles and Methods of Cartometry* (Oxford: Pergamon, 1989), 351–93.

40 "the simultaneous photographing . . . ": Tubis, "Aerial Photography," 23.

40 Coordination and training . . . : Procedures, operations, and developments discussed in this section are based on a variety of AAA documents as well as articles by USDA employees. Most noteworthy are Ralph Moyer, "Some Uses of Aerial Photographs in Connection with the Production and Marketing Programs of the U.S. Department of Agriculture," *Photogrammetric Engineering* 16 (1950): 305–7; Moyer, "Use of Aerial Photographs"; Department of Agriculture, Agricultural Adjustment Administration, North Central Region, *Procedure for Aerial Mapping in the State Office*, report NCR-State 104 (May 19, 1937); Department of Agriculture, Agricultural Adjustment Administration, Northeast Region, *County Procedure for Determination and Report of Performance (Applicable in Counties Using Aerial Photographs)*, report NER-329 (June 12, 1939); Department of Agriculture, Agricultural Adjustment Administration, Northeast Region, *Procedure for Determination and Report of Performance, Part II—Use of Aerial Photographs and Maps in Determining Performance*, report NER-219—Part II (June 20, 1938); Department of Agriculture, Agricultural Adjustment Administration, Southern Division, *Manual of Practice: Aerial Photography*, memorandum SRM-233 (November 23, 1938); Louis A. Woodward, "Aerial Photography as a Map Substitute," *Photogrammetric Engineering* 10 (1944): 68–81; and Marshall S. Wright, "The Aerial Photographic and

Photogrammetric Activities of the Federal Government," *Photogrammetric Engineering* 5 (1939): 168–76.

42 "in no event . . . ": Northeast Region, *County Procedure*, 7.

44 The relationship began . . . : Baldwin, Smith, and Whitlock, "Use of Aerial Photographs," 532–33.

44 When the United States . . . : Moyer, "Use of Aerial Photographs," 538.

44 For a concise history of military camouflage, see James F. Dunnigan and Albert A. Nofi, *Victory and Deceit: Dirty Tricks at War* (Fairfield, N.J.: William Morrow, 1995), 14–17. For an overview of infrared film, see Andrew Davidhazy, "Infrared Photography," in *The Focal Encyclopedia of Photography*, ed. Leslie Stroebel and Richard Zakia (Boston: Focal Press, 1993), 389–95.

45 "Natural grass . . . ": "Kodak Infrared Film Spots the 'Make Believe' of Enemy Camouflage" [advertisement], *United States News* 14 (March 12, 1943), 19.

46 Although most . . . : Page E. Truesdell, "Report of Unclassified Military Terrain Studies Section," *Photogrammetric Engineering* 19 (1953): 468–72.

47 For example, Landsat-7's . . . : See NASA, Earth Science Division, Ecosystem Science and Technology Branch, "Landsat 7," http://geo.arc.nasa .gov/sge/landsat/l7.html.

48 In addition to trouble-shooting . . . : Precision agriculture's unique cartographic genre includes the weed map, which recommends appropriately diverse applications of pesticide; for an example, see R. B. Brown and J.-P. G. A. Steckler, "Prescription Maps for Spatial Variable Herbicide Application in No-till Corn," *Transactions of the American Society of Agricultural Engineers* 38 (1995): 1659–66.

48 EC, which is . . . : N. R. Kitchen, K. A. Sudduth, and S. T. Drummond, "Soil Electrical Conductivity as a Crop Productivity Measure for Claypan Soils," *Journal of Production Agriculture* 12 (1999): 607–17; and E. D. Lund, C. D. Christy, and P. E. Drummond, "Applying Soil Electrical Conductivity Mapping to Improve the Economic Returns in Precision Farming" (paper presented at the 4th International Conference on Precision Agriculture, St. Paul, Minnesota, 1998).

48 "farming soils, not fields . . . ": P. M. Carr and others, "Farming Soils, Not Fields: A Strategy for Increasing Fertilizer Profitability," *Journal of Production Agriculture* 4 (1991): 57–61.

49 "It took more than . . . ": Pierre C. Robert, "Precision Agriculture: An Information Revolution in Agriculture" (paper presented at Agricultural Outlook Forum '98, Washington, D.C., February 23–24, 1999), 6.

49 For a survey of crop consultants' fees and activities, see R. J. Wright, T. A. DeVries, and S. T. Kamble, "Pest Management Practices of Crop Consultants in the Midwestern USA," *Journal of Production Agriculture* 10 (1997): 624–28.

49 For an intriguing description of one cutting crew's work and working conditions, see Shane DuBow, "Wheaties: Chasing the Ripening Harvest across America's Great Plains," *Harper's Magazine* 299 (August 1999): 33–44.

49 "the data will ... " : Dennis Berglund, "Precision Agriculture: Past, Present, Future" (paper presented at Agricultural Outlook Forum '98, Washington, D.C., February 23–24, 1999), 2.

50 Even so, authorities ... : "Scarecrow Points Way to Drugs for Police Officer in Helicopter," *New York Times*, June 30, 2000.

50 In the early 1990s ... : D. S. Fung and R. Welch, "Modeling Cannabis Cultivation in North Georgia," *Technical Papers of the American Congress on Surveying and Mapping/American Society of Photogrammetry and Remote Sensing Annual Convention*, 1994, 1:217–20.

50 For concise descriptions of forward-looking infrared (FLIR) imaging, see Abe Dane, "Night Hawks," *Popular Mechanics* 171 (November 1994): 78–81; Lillesand and Kiefer, *Remote Sensing and Image Interpretation*, 361–62; and William B. Scott, "Second-Generation FLIRs Enhance Night Attack Systems," *Aviation Week and Space Technology* 138 (June 7, 1993): 143–45. For an example of FLIR detection of an indoor "grow," see Tim Bryant, "DEA Targets Indoor Pot Growers," *St. Louis Post-Dispatch*, May 9, 1993.

50 Recent postings ... : See postings dated April 13 and 17, 2000, to the Security Forum of the Operation Overgrow bulletin board, http://www.hempcultivation.com/420/.

51 "undetected surveillance": Office of National Drug Control Policy, Counterdrug Technology Assessment Center, *Confronting Drug Crime and Abuse with Advanced Technology* (Washington, D.C., 2000), 13. The military used more sophisticated pilotless aircraft during the 2001 Afghanistan campaign; see James Dao, "U.S. Is Using More Drones, Despite Concern over Flaws," *New York Times*, November 3, 2001.

52 In rejecting Penny-Feeney's ... : *Unites States v. Penny-Feeney*, 773 F. Supp. 220 (D. Haw. 1991) at 225–26.

52 Judges in the Seventh ... : Douglas A. Kash, "Legal Development: Prewarrant Thermal Imaging as a Fourth Amendment Violation: A Supreme Court Question in the Making," *Albany Law Review* 60 (1997): 1295–1315; and NASA Office of Inspector General, "Remote Sensing and the Fourth Amendment: A New Law Enforcement Tool?" http://www.hq.nasa.gov/office/oig/hq/remote4.html.

53 "presumptively unreasonable": *United States v. Kyllo*, 140 F.3rd 1249 (9th Cir. 1998), quotation on 1253. Also see Bill Wallace, "Ninth Circuit Tosses Pot Conviction Case; Heat Detection Device Ruled Illegal," *San Francisco Chronicle*, April 8, 1998.

53 "Even if a thermal imager ... ": *United States v. Kyllo*, 1254–55.

53 "measured the heat ... ": Ibid., 1255.

53 "so revealing . . . ": *United States v. Kyllo,* 190 F.3rd 1041 (9th Cir. 1999), quotation on 1047; quotation is from the district court's original decision, which the appeals court now upheld.

53 "sense-enhancing . . . ": *Kyllo v. United States,* 150 L Ed 2d 94 (2001), quotation on 102.

53 "nothing more than . . . ": Ibid., 109.

53 "the device could not . . . ": Ibid., 112. Stevens included an appendix with examples of thermal images of Kyllo's home taken from a government exhibit presented at the trial; see ibid., 114.

53 "In the home . . . ": Ibid., 104.

53 "surprisingly broad": James J. Tomkovicz, quoted in Linda Greenhouse, "Justices Say Warrant Is Required in High-Tech Searches of Homes," *New York Times,* June 12, 2000.

54 The last time . . . : *Dow Chemical Company v. United States,* 106 S.Ct. 1819 (1986).

54 In ruling in favor . . . : Ibid., both quotations on 1823.

55 "the intimate activities . . . ": Ibid., 1826.

55 Moreover, Congress . . . : Ibid., 1827.

55 "highly sophisticated . . . ": Ibid.

55 After the *Wall Street Journal* . . . : Ross Kerber, "When Is a Satellite Photo an Unreasonable Search?" *Wall Street Journal,* January 27, 1998.

55 "high-tech military . . . ": The Libertarian Party memo "Do You Have Any Privacy Left When Big Brother Can Spy on You from Space—or through Your Walls" appeared on many Web sites, including the Canada Offshore Services, http://www.can-offshore.com/.

55 I doubt that many . . . : After the *Arizona Republic* reprinted Kerber's article, Pearson replied to his criticism in a letter to the editor; see Rita R. Pearson, "Tax-Payers Should Applaud High-Tech Enforcement of the Law," *Arizona Republic,* February 20, 1998.

56 Space Imaging no doubt . . . : Dan Leger, "Sizing Up the Competition: Earth Information Takes Commercial Intelligence to a New Level," *Imaging Notes* 15 (May/June 2000): 22–23. The courts might also need to rule on the vendor's liability when criminals, terrorists, or foreign governments use satellite imagery for illegal purposes. Would an imagery retailer be no more liable in these cases than, say, an Internet service provider or the phone company?

56 "highly sophisticated surveillance . . . ": *Dow Chemical Company v. United States,* 1827. For a review of terraserver.com, see Bill Siuru, "Spy Satellite Photography on the Internet," *Electronics Now* 70 (April 1999): 48–50.

Chapter 4. Tinder, Technology, and Tactics

General Sources

Basic references on the physics and chemistry of forest fuel and wildland fire include Margaret Fuller, *Forest Fires: An Introduction to Wildland Fire Behavior, Management, Firefighting, and Prevention* (New York: John Wiley and Sons, 1991), esp. 34–48, 69–80; and Stephen J. Pyne, Patricia L. Andrews, and Richard D. Laven, *Introduction to Wildland Fire*, 2nd ed. (New York: John Wiley and Sons, 1996), 3–168.

Sources on the use of GIS in modeling wildland fire include Vincent G. Ambrosia and others, "An Integration of Remote Sensing, GIS, and Information Distribution for Wildfire Detection and Management," *Photogrammetric Engineering and Remote Sensing* 64 (1998): 977–85; Maria J. Vasconcelos and D. Phillip Guertin, "FIREMAP—Simulation of Fire Growth with a Geographic Information System," *International Journal of Wildland Fire* 2 (1992): 87–96; and Rj Zimmer, "GIS and the Wildfires," *Professional Surveyor* 20 (December 2000): 55–59.

Principal sources on the use of satellite remote sensing in detecting wildland fire are Robert E. Burgan, Robert W. Klaver, and Jacqueline M. Klaver, "Fuel Models and Fire Potential from Satellite and Surface Observations," *International Journal of Wildland Fire* 8 (1998): 159–70; Donald R. Cahoon, Jr., and others, "Wildland Fire Detection from Space: Theory and Application," in *Biomass Burning and Its Inter-Relationships with the Climate System*, ed. John L. Innes, Martin Beniston, and Michel M. Verstraete (Dordrecht: Kluwer, 2000), 151–69; Emilio Chuvieco and M. Pilar Martin, "Global Fire Mapping and Fire Danger Estimation Using AVHRR Images," *Photogrammetric Engineering and Remote Sensing* 60 (1994): 563–70; Andrew J. L. Harris, "Towards Automated Fire Monitoring from Space: Semi-Automated Mapping of the January 1994 New South Wales Wildfires Using AVHRR Data," *International Journal of Wildland Fire* 6 (1996): 107–16; and Gary L. Hufford and others, "Use of Real-time Multisatellite and Radar Data to Support Forest Fire Management," *Weather and Forecasting* 13 (1998): 592–605.

Notes

57 For further information on the Cerro Grande Fire, see Department of the Interior, *Cerro Grande Prescribed Fire: Independent Review Board Report to the Secretary of the Interior*, May 26, 2000, http://www.doi.gov/secretary/reviewteamfinal.htm; and Michael Janofsky, "Parched U.S. Faces Worst Year for Fires since Mid-80s," *New York Times*, August 3, 2000.

58 For more information on the 1996 Mesa Verde National Park fire, see Joe Garner, "Mesa Verde Fire Tops $1.5 Million," *Denver Rocky Mountain News*, August 27, 1996; and Robert Kowalski, "Fire Damages Famed Petroglyph Stone Peeling at Mesa Verde Site," *Denver Post*, August 31, 1996.

58 Two weeks later . . . : "Colorado Fire Nears Old Indian Cliff Dwellings," *New York Times*, July 25, 2000; and Mindy Sink, "Scientists Unearth Artifacts a Step Ahead of Firefighters," *New York Times*, August 27, 2000.

59 During the 1930s, . . . : For examples, see *Fire Control Handbook, Section II: Presuppression* (n.p.: U.S. Forest Service, North Pacific Region, 1935), 7–7u; and L. G. Hornby, *Forest Fire Control Planning in the Northern Rocky Mountain Region* (n.p.: U.S. Forest Service, Rocky Mountain Forest and Range Experiment Station, 1936), 50–56.

60 Contemporary fuel maps . . . : For an example, see James K. Brown and Dennis G. Simmerman, *Appraising Fuels and Flammability in Western Aspen: A Prescribed Fire Guide*, USDA Forest Service Intermountain Research Station, General Technical Report Series, no. INT-205 (August 1986).

60 For further information on the National Fire Danger Rating System, see Pyne, Andrews, and Laven, *Introduction to Wildland Fire*, 155–59. For a description of the statistical rationale and computer program, see Patricia L. Andrews and Larry S. Bradshaw, *FIRES: Fire Information Retrieval and Evaluation System—A Program for Fire Danger Rating Analysis*, USDA Forest Service Intermountain Research Station, General Technical Report Series, no. INT-367 (August 1997).

62 "the fires literally . . . ": Quoted in Michael Paterniti, "Torched," *Outside* 20 (September 1995): 57–68, 154–56; quotation on 66.

62 For details on early fire-spread models, see Richard C. Rothermel, *How to Predict the Spread and Intensity of Forest and Range Fires*, USDA Forest Service Intermountain Forest and Range Experiment Station, General Technical Report Series, no. INT-143 (June 1983).

63 "upon the skill . . . ": Ibid., 1.

64 For a concise account of California's use of GIS in controlling wildland fire, see Tim Walsh, "In the Line of Fire," *Geospatial Solutions* 10 (June 2000): 24–29.

64 The fire burned . . . : Data are from Sarah G. Allen and others, "Interactive Application of GIS During the Vision Wildfire at Point Reyes National Seashore" (paper presented at the 1996 Arc/Info Users Conference, Palm Springs, Cal., May 20–24, 1996), National Park Service, http://www.nps.gov/gis/apps/pore/gisndx.htm.

65 "great tools for public relations": Quoted in the minutes of the National Park Service GIS Workshop, George Wright Society Meeting, Albuquerque, NM, March 20, 1997, National Park Service, http://www.nps.gov/gis/education/gwsociety_notes.html.

65 Back in the 1980s . . . : Jay Lee, "Analyses of Visibility Sites on Topographic Surfaces," *International Journal of Geographical Information Systems* 5 (1991): 413–29.

65 For further information on POES, see Arthur P. Cracknell, *The Advanced*

Very High Resolution Radiometer (AVHRR) (London: Taylor and Francis, 1997), 5–26, 36–43.

65 For further information on GOES, see W. Paul Menzel and James F. W. Purdom, "Introducing GOES-I: The First of a New Generation of Geostationary Operational Environmental Satellites," *Bulletin of the American Meteorological Society* 75 (1994): 757–81.

66 This markedly higher orbit . . . : At 60 degrees from the equator resolution falls to approximately 5 miles (8 km). See Gary L. Hufford and others, "Detection and Growth of an Alaskan Forest Fire Using GOES-9 3.9 m Imagery," *International Journal of Wildland Fire* 9 (1999): 129–36.

66 My discussion of the Millers Reach fire is based on Hufford and others, "Use of Real-time Multisatellite Data."

68 "an ultimate detection time . . . ": "Space Imaging Techniques and Fire Management (Interim Report)," NOAA CEOS Disaster Information Server, http://www.ceos.noaa.gov/drafts/firerpt.html (site discontinued).

68 Equally intriguing . . . : Joel S. Levine and others, "FireSat and the Global Monitoring of Biomass Burning," in *Biomass Burning and Global Change*, vol. 1, *Remote Sensing, Modeling and Inventory Development, and Biomass Burning in Africa*, ed. Joel S. Levine (Cambridge, Mass.: MIT Press, 1996), 107–29; and NASA Langley Research Center, "1998 Langley Research Center Implementation Plan, Section II: Scientific Roles in Support of Earth Science Enterprise,"http://larcip.larc.nasa.gov/1998/section2.html.

68 According to research . . . : See, for example, Cahoon and others, "Wildland Fire Detection."

69 "a constellation of sixty satellites": Joel S. Levine, telephone conversation with author, October 4, 2000.

70 Another element is GOES . . . : Burgan, Klaver, and Klaver, "Fuel Models and Fire Potential"; and Dennis S. Mileti, *Disasters by Design: A Reassessment of Natural Hazards in the United States* (Washington, D.C.: Joseph Henry Press, 1999), 181. For a modified, May 2000 version of the paper by Burgan, Klaver, and Klaver, see the USDA Forest Service, http://www.fs.fed.us/land/wfas/firepot/fpipap.htm.

70 As in camouflage detection . . . : Relative greenness is based on the Normalized Difference Vegetation Index (NDVI), which compares reflectance measurements in visible red and near-infrared AVHRR bands observed for the pixel over a two-week period. Relative greenness compares the pixel's current NDVI to the maximum and minimum NDVI values observed since January 1, 1989. For formulas and a fuller explanation, see Robert E. Burgan and Roberta A. Hartford, *Monitoring Vegetation Greenness with Satellite Data*, USDA Forest Service Intermountain Research Station, General Technical Report Series, no. INT-297 (May 1993).

70 For a description of the system as well as current and archived maps, see

the Oklahoma Fire Danger Model, http://radar.metr.ou.edu/agwx/fire/. For a concise description of Mesonet and efforts to make younger residents aware of weather hazards, see Renee A. McPherson and Kenneth C. Crawford, "The Earthstorm Project: Encouraging the Use of Real-Time Data from the Oklahoma Mesonet in K-12 Classrooms," *Bulletin of the American Meteorological Society* 77 (1996): 749–61.

72 Equally inefficient . . . : Andrew C. Revkin, "Now Preventing Forest Fires: Smokey Goat," *New York Times*, June 13, 2000.

Chapter 5. Weather Eyes

General Sources

Key sources on satellite weather surveillance include W. Paul Menzel and others, "Application of GOES-8/9 Soundings to Weather Forecasting and Nowcasting," *Bulletin of the American Meteorological Society* 79 (1998): 2059–77; W. Paul Menzel and James F. W. Purdom, "Introducing GOES-I: The First of a New Generation of Geostationary Operational Environmental Satellites," *Bulletin of the American Meteorological Society* 75 (1995): 757–81; and Space Systems-Loral, *GOES I-M DataBook* (August 31, 1996), NASA Goddard Space Flight Center, http://rsd.gsfc.nasa.gov/goes/text/goes.databook.html.

Sources on the development of forecast models include Mark DeMaria, "A History of Hurricane Forecasting for the Atlantic Basin, 1920–1995," in *Historical Essays on Meteorology, 1919–1995*, ed. James Rodger Fleming (Boston: American Meteorological Society, 1996), 263–305; and Colin J. McAdie and Miles B. Lawrence, "Improvements in Tropical Cyclone Track Forecasting in the Atlantic Basin, 1970–98," *Bulletin of the American Meteorological Society* 81 (2000): 989–97.

Useful references on the development of weather radar include Elbert W. Friday, Jr., "The Modernization and Associated Restructuring of the National Weather Service: An Overview," *Bulletin of the American Meteorological Society* 75 (1994): 43–52; and Robert J. Serafin and James W. Wilson, "Operational Weather Radar in the United States: Progress and Opportunity," *Bulletin of the American Meteorological Society* 81 (2000): 501–18.

Key sources for lightning detection are Hugh J. Christian and others, "Lightning Detection from Space (A Lightning Primer)," NASA Global Hydrology and Climate Center, http://wwwghcc.msfc.nasa.gov/lisotd_old.html; K. L. Cummins and others, "A Combined TOA/MDF Technology Upgrade of the US National Lightning Detection Network," *Journal of Geophysical Research D: Atmospheres* 103 (April 27, 1998): 9035–44; NOAA National Weather Service Training Center, "Lightning Detection Systems," http://www.nwstc.noaa.gov/d.MET/Lightning/detection.htm; and Richard E. Orville, Ronald W. Henderson, and Lance F. Bosart, "An East Coast Lightning Detection Network," *Bulletin of the American Meteorological Society* 64 (1983): 1029–37.

Notes

74   In *Isaac's Storm* . . . : Erik Larson, *Isaac's Storm: A Man, a Time, and the Deadliest Hurricane in History* (New York: Random House, 1999).

74   Turn-of-the-century forecasters . . . : Mark Monmonier, *Air Apparent: How Meteorologists Learned to Map, Predict, and Dramatize Weather* (Chicago: University of Chicago Press, 1999), 10–15.

75   The antagonist . . . : Sebastian Junger, *The Perfect Storm* (New York: W. W. Norton, 1997).

75   This one, later named . . . : NOAA Marine Prediction Center, "The Marine Prediction Center and 'The Perfect Storm,'" http:www.mpc.ncep.noaa .gov/perfectstorm/mpc_ps_intro.html; and NOAA National Climatic Data Center, "The Perfect Storm," http://www.ncdc.noaa.gov/ol/satellite/ satelliteeye/cyclones/pfctstorm91/pfctstorm.html.

75   For a summary of the effects of the Halloween Storm of 1991, see Robert D. McFadden, "Report on Damage in Storm Is Grim," *New York Times*, November 3, 1991, sec. 1.

76   American and Canadian . . . : NOAA National Climatic Data Center, "Unnamed Hurricane, 1991," http://www.ncdc.noaa.gov/ol/satellite/ satelliteeye/hurricanes/unnamed91/unnamed91.html.

76   As hurricanes go . . . : Ed Rappaport, "Hurricane Andrew, 16–28 August 1992: Preliminary Report (updated 10 December 1993)," NOAA National Hurricane Center (Tropical Prediction Center), http://www.nhc.noaa .gov/1992andrew/html.

77   According to Louis Uccellini . . . : "NOAA Meteorologists Recall Drama of Forecasting 'The Perfect Storm,'" *NOAA Magazine,* June 29, 2000, http://www.noaanews.noaa.gov/stories/s451.htm.

78   "if there was a choice . . . ": Robert C. Sheets, "The National Hurricane Center—Past, Present, and Future," *Weather and Forecasting* 5 (1990): 185–232; quotation on 201.

78   Although GOES satellites . . . : U.S. General Accounting Office, *National Oceanic and Atmospheric Administration: National Weather Service Modernization and Weather Satellite Program* (statement of Joel C. Willemssen), report no. T-AIMD-00-86 (March 29, 2000).

79   Three imaging schedules . . . : See NOAA Office of Satellite Operations, "GOES Schedules and Scan Sectors," http://www.oso.noaa.gov/goes/ schd-sector/index.htm.

80   Able to scan . . . : According to Daphne S. Zaras at the National Severe Storms Laboratory (NSSL), "Super Rapid Scan is difficult to describe. The time interval between images is extremely irregular, in order to maintain

scanning of the rest of the CONUS (during each hour) and full disk (every 3 hours) while doing shorter interval scanning over a 'mesoscale' (1 km × 1 km) area. The size of the super rapid scan area is conveniently the size of the area affected by the 'Super Outbreak' of tornados April 3–4, 1974." Zaras, "GOES-IM Weather Satellites," NOAA National Severe Storms Center, http://www.nssl.noaa.gov/~zaras/Goes/delux2/scan3.html.

81 As a result, hurricane . . . : AMS Council, "Policy Statement on Hurricane Research and Forecasting," *Bulletin of the American Meteorological Society* 81 (2000): 1341–46.

81 For information on the Tropical Rainfall Measuring Mission, see NASA's TRMM Web site, http://trmm.gsfc.nasa.gov/.

82 Among the TRMM . . . : Riko Oki, Kinji Furukawa, and Shuji Shimizu, "Preliminary Results of TRMM: Part I, a Comparison of PR with Ground Observations," *Marine Technology Science Journal* 32 (winter 1998–99): 13–23.

82 This map gained . . . : William Booth and Mary Jordan, "Hurricane Rips Miami Area, Aims at Gulf States," *Washington Post,* August 25, 1992.

82 The following April . . . : Linda Martin, "National Weather Service Gets Advanced Doppler Radar," *Miami Herald,* April 15, 1993.

84 For example, a 1996 . . . : J. Madeleine Nash, "Unraveling the Mysteries of Twisters," *Time* 147 (May 20, 1996): 58–62; and National Science and Technology Council, Committee on Environment and Natural Resources, Subcommittee on Natural Disaster Reduction (SNDR), "Agency Success Stories in Natural Disaster Reduction," http://www.usgs.gov/sndr/success.html.

84 By contrast, a NOAA . . . : Curtis L. Marshall, "Strategic Planning in the National Weather Service: Case Study (June 9, 1997)," National Partnership for Reinventing Government, http://govinfo.library.unt.edu/npr/library/news/ntlwpln.html; and NOAA National Weather Service, "Reinvention Goals for 2000: Status—August 2000," http://www.nws.noaa.gov/npr5.html.

84 These and other claims . . . : Question 0200 of the WeatherQuiz, KRT Direct, April 8, 1999, http://www.krtdirect.com/weatherquiz/.

85 Because radar maps . . . : See, for example, Caren Marzban and Gregory J. Stumpf, "A Neural Network for Tornado Prediction Based on Doppler Radar-Derived Attributes," *Journal of Applied Meteorology* 35 (1996): 617–26.

85 Developed, tested . . . : "Reinvention Goals for 2000."

85 At a June 1999 hearing . . . : House Committee on Science, Subcommittee on Energy and Environment, *Tornadoes: Understanding, Modeling, and Forecasting Supercell Storms—Hearing before the Subcommittee on Energy and Environment and the Subcommittee on Basic Research,* 106th Cong., 1st sess., June 16, 1999; quotation on 37.

85 "Doppler radars . . . ": Ibid., 38.

85 Most of Oklahoma's . . . : The 180-kilometer range is a median distance reflecting the average experience of forecasters and researchers. NEXRAD Panel, National Weather Service Modernization Committee, Commission on Engineering and Technical Systems, National Research Council, *Toward a New National Weather Service: Assessment of NEXRAD Coverage and Associated Weather Services* (Washington, D.C.: National Academy Press, 1995), 12–17.

86 "a challenging event . . . ": Quoted in Jessica Gregg McNew, "Record Tornado Season Has Forecasters, Scientists Gathering for Answers," *DisasterRelief*, August 19, 1998, http://www.disasterrelief.org/Disasters/980819forum/. For additional information about the storm, see NOAA National Weather Service, Southern Region, *The Hall/White County Tornado, March 20, 1998,* National Weather Service Southern Region Service Assessment Report (May 1998), http://www.srh.noaa.gov/ftproot/msd/html/assessment/hallwhit.htm.

86 According to a 1995 . . . : NEXRAD Panel, *Toward a New National Weather Service,* 16.

86 Moreover, the 15 percent . . . : Ibid., 21–23.

87 At thirty locations . . . : Edward J. Hopkins, "Wind Profiler," University of Wisconsin Department of Meteorology, http://www.meteor.wisc.edu/~hopkins/aos100/wxi-prfl.htm; and the NOAA Profiler Network pages on the Forecast Systems Laboratory's Demonstration Division Web site, http://www.profiler.noaa.gov/jsp/index.jsp.

88 "the computer models . . . ": Quoted in Margot Ackley, Douglas W. van de Kamp, and Seth I. Gutman, "Profiling," NOAA Forecast Systems Laboratory, Demonstration Division, http://www.profiler.noaa.gov/labreview/1999/sld023.htm.

89 Although the Wind Profilers . . . : The Wind Profiler network will become officially "operational" when NOAA's research arm converts all profilers to an exclusive, authorized frequency, and the National Weather Service assumes operating responsibility. Margot Ackley, NOAA Forecast Systems Laboratory, telephone conversation with author, November 28, 2000.

90 To get a sharper picture . . . : See the NOAA National Weather Service Training Center Web site, http://www.nwstc.noaa.gov/d.MET/Lightning/Forecast_use.htm.

91 For information on NOAA's Ground-Based GPS Integrated Precipitable Water Vapor Demonstration Network, see the NOAA Forecast Systems Laboratory Web site, http://www.gpsmet.noaa.gov/jsp/index.jsp; and Steven Businger and others, "The Promise of GPS in Atmospheric Monitoring," *Bulletin of the American Meteorological Society* 77 (1996): 5–18.

91 For a sampling of the trials and triumphs of AWIPS, see U.S. General Ac-

counting Office, *Weather Forecasting: NWS Has Not Demonstrated That New Processing System Will Improve Mission Effectiveness,* report no. AIMD-96-29 (February 29, 1996); Jim Reed, "Weather's New Outlook," *Popular Science* 252 (August 1997): 56–62; and "NOAA Installs New Weather Forecasting System," *Bulletin of the American Meteorological Society* 80 (1999): 2121–22.

Chapter 6. Wire Loops and Traffic Cams

General Sources

Principal sources for traffic engineering are John E. Baerwald, ed., *Traffic Engineering Handbook,* 3rd ed. (Washington, D.C.: Institute of Traffic Engineers, 1965); John D. Edwards, Jr., ed., *Transportation Planning Handbook,* 2nd ed. (Washington, D.C.: Institute of Transportation Engineers, 1999); and William R. McShane and Roger P. Roess, *Traffic Engineering* (Englewood Cliffs, N.J.: Prentice-Hall, 1990).

Key references for video-controlled traffic signals are Carlo S. Regazzoni, Gianni Fabri, and Gianni Vernazza, eds., *Advanced Video-Based Surveillance Systems* (Boston: Kluwer Academic Publishers, 1999); Panos G. Michalopoulos, "Vehicle Detection Video Through Image Processing: The Autoscope System," *IEEE Transactions on Vehicular Technology* 40 (1991): 21–29; and Panos G. Michalopoulos and Kevin Samartin, "Recent Developments of Advanced Technology in Freeway Management Projects," *Traffic Engineering and Control* 39 (1998): 160–65.

Sources for electronic toll collection include Lazar N. Spasovic and others, "Primer on Electronic Toll Collection Technologies," *Transportation Research Record,* no. 1516 (Washington, D.C.: National Academy Press, 1995), 1–10; and ETTM On the Web, which provides detailed information on E-ZPass, including Michael Kolb's detailed and richly illustrated essay, "ETC in Focus: The Port Authority of NY and NJ," http://www.ettm.com/focus_pa/etcfocus_panynj.html.

For insights on automated driverless transportation, I relied on Steven Ashley, "Smart Cars and Automated Highways," *Mechanical Engineering* 120 (May 1998) 58–62; Glen Hiemstra, "Driving in 2020: Commuting Meets Computing," *Futurist* 34 (September/October 2000): 31–34; James H. Rillings, "Automated Highways," *Scientific American* 277 (October 1997): 80–85; and Horst Strobel, *Computer Controlled Urban Transportation* (Chichester, U.K.: John Wiley and Sons, 1982), esp. 355–411. For diverse views of the future of automated roadways, I recommend José M. del Castillo, David J. Lovell, and Carlos F. Daganzo, "Technical and Economic Viability of Automated Highway Systems: Preliminary Analysis," *Transportation Research Record,* no. 1588 (Washington, D.C.: National Academy Press, 1997), 130–36; and Steven F. Shladover, "Why We Should Develop a Truly Automated Highway System," *Transportation Research Record,* no. 1651 (Washington, D.C.: National Academy Press, 1998), 66–73.

Notes

94 For additional information about "road tubes" and other techniques for measuring traffic volume, see James H. Kell and Wolfgang S. Homburger, "Traffic Studies," in *Traffic Engineering Handbook*, 260–309, esp. 268–72. Oddly, the fifth edition of the *Handbook*, published in 2000, omits discussion of road tubes and other sensors, which are still widely used for traffic studies.

94 For guidelines used in computing splits and setting phases for isolated or nonsynchronized signals, see Herbert J. Klar, "Traffic Signalization," in *Traffic Engineering Handbook*, 382–458, esp. 404–10; and McShane and Roess, *Traffic Engineering*, 380–413.

94 For examples of time-space diagrams in traffic-signal synchronization, see Klar, "Traffic Signalization," 413–33; McShane and Roess, *Traffic Engineering*, 529–48; Peter S. Parsonson, *Signal Timing Improvement Practices*, National Cooperative Highway Research Program, Synthesis of Highway Practice report no. 172 (Washington, D.C.: Transportation Research Board, 1992), 31–45; and Jearl Walker, "How to Analyze a City Traffic-Light System from the Outside Looking In," *Scientific American* 248 (March 1983): 138–45.

96 For examples of "double-alternate" timing and other synchronization strategies for two-way traffic, see McShane and Roess, *Traffic Engineering*, 543–48.

96 For information on ramp-metering strategies for freeways, see ibid., 617–32.

97 Although the detector . . . : See Gaylon R. Claiborne, "Induction Vehicle Detectors for Traffic Actuated Signals," *Traffic Engineering* 33 (December 1962): 21–25; and Indu Sreedevi and Justin Black, "Inductive Loop Detectors," University of California, Berkeley, Partners for Advanced Transit and Highways ITS Decision Database, http://www.path.Berkeley.edu/~leap/TTM/Incident_Manage/Detection/loopdet.html.

97 For detailed information on the underlying physics, design, installation, and performance of loop detectors, see James H. Kell, Iris J. Fullerton, and Milton K. Mills, *Traffic Detector Handbook*, 2nd ed., Federal Highway Administration report no. FHWA-IP-90-002 (Washington, D.C., 1990), esp. 7–15, 105–12, and 133–34. The handbook is available at the Federal Highway Administration Safety and Operations Electronic Reading Room, http://www.fhwa.dot.gov/tfhrc/safety/pubs/Ip9000/intro.htm.

98 For discussion of the relative advantages of various shapes of loop detectors, see ibid., 73–79; and Sreedevi and Black, "Inductive Loop Detectors."

99 Reliable counts are . . . : For an overview of traffic counting in Delaware and a map of detector locations, see Ardeshir Faghri, Martin Glaubitz, and Janaki Parameswaran, "Development of Integrated Traffic Monitoring

System for Delaware," *Transportation Research Record*, no. 1536 (Washington, D.C.: National Academy Press, 1996), 40–44.

99 Toronto and other . . . : "Stop on the Dots," *Cyclometer*, no. 61 (May—June 1997), http://www.city.scarborough.on.ca/4service/cyclom61.htm (site discontinued).

100 For discussion of detector-based strategies to lower the likelihood of catching vehicles in the dilemma zone, see Kell, Fullerton, and Mills, *Traffic Detector Handbook*, 70–72, 81–88.

100 In the Texas Panhandle . . . : Beth Wilson, "Eye in the Sky: Camera Sees All, Clicks All," *Amarillo Globe-News*, March 4, 2000; and David Miller, Texas Department of Transportation, Amarillo District, telephone conversation with the author, January 16, 2001.

100 For additional information on the principles of machine vision, see Bin Ran and Henry X. Liu, "Development of Vision-Based Vehicle Detection and Recognition System for Intelligent Vehicles," *Transportation Research Record*, no. 1679 (Washington, D.C.: National Academy Press, 1999), 130–38; and Mark R. Stevens and J. Ross Beveridge, *Integrating Graphics and Vision for Object Recognition* (Boston: Kluwer Academic Publishers, 2000).

101 Not all traffic . . . : This section is based on my January 17, 2001, interview with traffic control room operator Harry Carlson and Luis Perez, "Synchronized Stopping: Computers Control Traffic Lights in Downtown Syracuse," *Syracuse Post-Standard*, May 5, 2000.

102 MIST is one of several traffic control software applications sold to state and local transportation departments. The developer, PB Farradyne, provides further information on its Web site, http://www.pbfi.com. For a general assessment of traffic control software, see Darcy Bullock and Chris Hendrickson, "Advanced Software Design and Standards for Traffic Signal Control," *Journal of Transportation Engineering* 118 (May/June 1992): 430–38.

102 For an insightful examination of the role of traffic signaling in improving air quality, see Committee for the Study of Impacts of Highway Capacity Improvements on Air Quality and Energy Consumption, *Expanding Metropolitan Highways: Implications for Air Quality and Energy Use*, Transportation Research Board Special Report 245 (Washington, D.C.: National Academy Press, 1995), esp. 38–86.

103 Motorists curious . . . : Counts are based on index maps displayed on January 24, 2001. For a fuller discussion of Internet delivery of traffic information, see Nicoline N. M. Emmer, "Web Maps and Road Traffic," in *Web Cartography: Developments and Prospects*, ed. Menno-Jan Kraak and Allan Brown (London: Taylor and Francis, 2001), 159–70; and Glenn D. Lyons and Mike McDonald, "Traveller Information and the Internet," *Traffic Engineering and Control* 39 (1998): 24–31.

104   Each icon is . . . : For further discussion of the relationship between web cams and interactive maps, see Mark Monmonier, "Webcams, Interactive Index Maps, and Our Brave New World's Brave New Globe," *Cartographic Perspectives* no. 37 (fall 2000): 12–25.

105   Washington transportation . . . : David Noack, "Puget Sound's Web Cam Traffic Network: DOT Offers Online Traffic News" (February 6, 1998), Editor and Publisher's interactive media Web site [now http://www.editorandpublisher.com], http://www.mediainfo.com/ephome/Interactive 98/stories/020698c9.htm. Also see Washington State Department of Transportation Puget Sound Traffic Conditions, "Questions and Answers," http:www.wsdot.wa.gov/PugetSoundTraffic/faq/.

105   Web traffic increased . . . : George Foster, "For Some Web Sites, a Lot of Traffic Means Lots of Online Traffic," *Seattle Post-Intelligencer,* December 19, 1999.

105   Tacoma's city-run . . . : Martha Modeen, "TV Carries Scenes from DOT Traffic Cameras," *Tacoma News Tribune,* November 2, 2000.

106   Named Partners in Motion . . . : See Marcia Myers, "Region's Road Warriors Battle-Test New Weapon," *Baltimore Sun,* January 4, 2000; and Alice Reid, "High-Tech Traffic Help Is En Route," *Washington Post,* August 10, 1998. Web sites offering the traffic reports include *NBC4.com* and *washingtonpost.com* as well as the firm's own SmarTraveler.com.

106   Unlike Puget Sound's . . . : For example, the "live video" of the Woodrow Wilson Bridge presents a mildly jerky "streaming video" image in a RealPlayer viewer, which is refreshed every two minutes with a new file encoded at 15 frames per second but displayed at a lower rate. See "Wilson Bridge Live Video," *washingtonpost.com,* http://www.washingtonpost.com/wp-srv/local/traffic/wash1.htm.

106   The number of traffic cameras in the greater Washington area seems likely to increase. As of January 2000, Virginia's Department of Transportation planned to operate 110 cameras by June, its Maryland counterpart had 42 cameras statewide, and Montgomery County, Maryland, planned to add to its 83 cameras. See Leslie Koren, "New VDOT Website Shows Drivers the Way," *Washington Times,* January 6, 2000.

106   In addition to . . . : For examples, see Paul Bradley, "Online Road Views Aim to Ease Commute," *Richmond Times-Dispatch,* January 6, 2000; Diane Granat, "Traffic Busters," *Washingtonian* 34 (September 1999): 86–93; Joey Ledford, "'Personalized Traffic'—How to Avoid the Rush," *Atlanta Constitution,* September 15, 1999; and Bill Steward, "High Tech Employed to Outfox Snarls," *The Oregonian,* June 19, 2000. Also see Beth Cox, "Wireless Highway Data Services Planned," *InternetNews—E-Commerce News,* May 3, 2000, http://www.internetnews.com/ec-news/article/0,,4_353451,00.html. Microsoft introduced a similar service for Seattle drivers; see Brier Dudley, "Microsoft Offers Drivers High-Tech

Traffic Alerts," *seattletimes.com*, October 24, 2001, http://seattletimes
.nwsource.com/html/businesstechnology/134354442_msn16.html.

106 A camera mounted . . . : For examples, see Stacey Burns, "Run a Light?
Ticket's in the Mail," *Tacoma News Tribune*, August 24, 2000; and Marty
Katz, "Frown, You're on Red-Light Camera," *New York Times*, October 11,
2000.

107 Most states . . . : Jennifer Jones, "Market for Red-Light-Running Systems
Speeds Up" (June 7, 1999), *FCW.COM*, http://www.civic.com/civic/
articles/1999/CIVIC_060799_57.asp. For a concise overview of digital
license plate readers, see Catherine Greenman, "Zeroing in on the Suspicious Number above the State Motto," *New York Times*, October 25,
2001.

107 . . . an alarming increase . . . : Richard A. Retting, Robert G. Ulmer, and
Allan F. Williams, "Prevalence and Characteristics of Red Light Running
Crashes in the United States," *Accident Analysis and Prevention* 31 (1999):
687–94.

107 Studies indicate . . . : Richard A. Retting and Allan F. Williams, "Red Light
Cameras and the Perceived Risk of Being Ticketed," *Traffic Engineering
and Control* 41 (2000): 224–27.

108 "What's going on . . . ": Quoted in Aron Miller, "Use of Stoplight Cameras
Causes Swirl of Debate," *Ventura County Star*, May 30, 2000.

108 "general surveillance . . . ": Quoted in William Claiborne, "California Assembly Puts Stop to Red Light Cameras," *Washington Post*, May 10, 1998.

108 "It's not a new . . . ": Claire E. House, "Legislatures Debate Merits of Stoplight Photos" (July 1999), *Government Computer News*, http://www.gcn
.com/state/vol5_no7/news/375–1.html.

108 My hunch . . . : Portland, Oregon, for example, won legislative approval
after officials agreed to a camera that only photographs violators. See
Associated Press, "Portland Keeps Its Eye on Red Lights," *Seattle Post-
Intelligencer*, January 22, 1999.

108 A vehicle approaching . . . : For general information on this kind of
transponder, see Jian John Lu, Michael J. Rechtorik, and Shiyu Yang, "Automatic Vehicle Identification Technology Applications to Toll Collection
Services," *Transportation Research Record*, no. 1588 (Washington, D.C.: National Academy Press, 1997), 18–25.

108 Market-based . . . : For a glimpse of this pay-as-you-go future, see Richard
H. M. Emmerink, *Information and Pricing in Road Transportation* (Berlin:
Springer, 1998); and Charles Komanoff and Michael J. Smith, "It Isn't Too
Many Double-Parkers; It's Too Many Cars," *New York Times*, October 16,
1999. Another likely proving ground is Singapore, a conveniently com-
pact country that is familiar with regimentation; see A. P. Gopinath
Menon, "ERP in Singapore—a Perspective One Year On," *Traffic Engi-
neering and Control* 41 (February 2000): 40–45. Although road pricing is

unlikely to affect North America in the near future, Europe is ahead in other approaches to electronic traffic control as well. For examples, see Linda L. Brown and others, "Innovative Traffic Control: Technology and Practice in Europe—Executive Summary," *ITE Journal* 70 (2000): 45–49.

109 Bolstered by GPS . . . : For examples, see Bill McGarigle, "Full Stop," *Government Technology* 14 (May 2000): 50–54.

109 "Smart cars . . . ": For information about collision avoidance systems, see Sandy Graham, "Smart Cars Open Way for Safer, Faster Travel," *Traffic Safety* (March/April 2000): 17–19; and Peter Godwin, "The Car That Can't Crash," *New York Times Magazine*, June 11, 2000.

109 Equally certain . . . : For further insight on privacy concerns, see Sheri Alpert and Kingsley E. Haynes, "Privacy and the Intersection of Geographical Information and Intelligent Transportation Systems," in *Proceedings of the Conference on Law and Information Policy for Spatial Databases*, ed. Harlan J. Onsrud (Orono, Maine: National Center for Geographic Information and Analysis, University of Maine, 1995), 198–211.

Chapter 7. Crime Watch

General Sources

Central references on video surveillance include David Lyon, *The Electronic Eye: The Rise of Surveillance Society* (Minneapolis: University of Minnesota Press, 1994), 67; Clive Norris, Jade Moran, and Gary Armstrong, eds., *Surveillance, Closed Circuit Television, and Social Control* (Aldershot, U.K.: Ashgate, 1998); and Carlo S. Regazzoni, Gianni Fabri, and Gianni Vernazza, eds., *Advanced Video-Based Surveillance Systems* (Boston: Kluwer, 1999), 95–105.

General sources covering crime mapping include Arthur Getis and others, "Geographic Information Science and Crime Analysis," *URISA Journal* 12 (spring 2000): 7–14; Keith Harries, *Mapping Crime: Principle and Practice*, research report NCJ 178919 (Washington, D.C.: National Institute of Justice, 1999); Jack Maple and Chris Mitchell, *The Crime Fighter: Putting the Bad Guys Out of Business* (New York: Doubleday, 1999); Phillip D. Phillips, "A Prologue to the Geography of Crime," *Proceedings of the Association of American Geographers* 4 (1972): 86–91; and Arthur H. Robinson, *Early Thematic Mapping in the History of Cartography* (Chicago: University of Chicago Press, 1982), 156–70.

Notes

110 In 1791, he published . . . : Jeremy Bentham, *"Panopticon": or, the Inspection-House; containing the idea of a new principle of construction applicable to any sort of establishment, in which persons of any description are to be kept under inspection; and in particular to Penitentiary-houses, Prisons, Houses of in-*

dustry, Workhouses, Poor Houses, Manufacturies, Madhouses, Lazarettos, Hospitals, and Schools; with a plan of management adopted [sic] *to the principle; in a series of letters, written in the year 1787, from Crechoff in White Russia, to a friend in England* (London: T. Payne, 1791). Note that the drawing in figure 7.1 is from Bentham's collected writings, published posthumously by his literary executor.

111   "panoptic gaze" . . . "panoptic power": For examples, see Arturo Escobar, *Encountering Development: The Making and Unmaking of the Third World* (Princeton, N.J.: Princeton University Press, 1995), 155–56; N. Katherine Hayles, "The Materiality of Informatics," *Configurations* 1 (1993): 147–70, esp. 150–52; and Lyon, *Electronic Eye,* 67. As numerous authors point out, commercial interests as well as governments seek social control, albeit for different reasons and usually in different ways. The concept of the panoptic gaze is generally attributed to Michel Foucault; see Foucault, *Discipline and Punish: The Birth of the Prison,* trans. Alan Sheridan (New York: Pantheon, 1978; Vintage Books, 1995), 200–216.

112   For a sense of Britain's commitment to video surveillance, see Jason Ditton and Emma Short, "Evaluating Scotland's First Town Centre CCTV Scheme," in *Surveillance, Closed Circuit Television, and Social Control,* ed. Norris, Moran, and Armstrong, 155–73; and Nicholas R. Fyfe and Jon Bannister, "City Watching: Closed Circuit Television Surveillance in Public Spaces," *Area* 28 (1996): 37–46.

112   A survey by the California Research Bureau . . . : Marcus Nieto, *Public Video Surveillance: Is It an Effective Crime Prevention Tool?* report no. CRB-97-005 (Sacramento, Cal.: California Research Bureau, California State Library, 1997).

113   Baltimore, for instance . . . : A 1997 *New York Times* article labeled Baltimore's setup "one of the country's most ambitious video surveillance programs." See Michael Cooper, "With Success of Cameras, Concerns over Privacy," *New York Times,* February 5, 1997.

113   "if we start going . . .": Quoted in David Kocieniewski, "Television Cameras May Survey Public Places," *New York Times,* October 6, 1996.

113   "raises the Orwellian specter . . .": Quoted in David Kocieniewski, "Police to Press Property-Crime Fight and Install Cameras," *New York Times,* February 5, 1997.

114   Intent on documenting . . . : See New York Civil Liberties Union, "NYCLU Surveillance Camera Project," http://www.nyclu.org/surveillance.html. For the map, a more detailed explanation, and an interpretation, see Mediaeater, "NYC Surveillance Camera Project," http://www.mediaeater.com/cameras/.

115   For examples of social science critiques of CCTV, see William Bogard, *The Simulation of Surveillance: Hypercontrol in Telematic Societies* (Cambridge: Cambridge University Press, 1996); and Stephen Graham, "Spaces of

Surveillant Simulation: New Technologies, Digital Representations, and Material Geographies," *Environment and Planning D: Society and Space* 16 (1998): 483–504.

115 Evaluation studies are . . . : For insights, see David Skinns, "Crime Reduction, Diffusion, and Displacement: Evaluating the Effectiveness of CCTV," in *Surveillance, Closed Circuit Television, and Social Control*, ed. Norris, Moran, and Armstrong, 175–88; and Nick Tilley, "Evaluating the Effectiveness of CCTV Schemes," in *Surveillance, Closed Circuit Television, and Social Control*, ed. Norris, Moran, and Armstrong, 139–53.

115 . . . a majority of Britons . . . : Although how one phrases the question can have a substantial effect on the percentage of respondents with a favorable impression of CCTV, those who support the technology are a clear majority. For an examination of opinion surveys, see Jason Ditton, "Public Support for Town Centre CCTV Schemes: Myth or Reality?" in *Surveillance, Closed Circuit Television, and Social Control*, ed. Norris, Moran, and Armstrong, 221–28.

115 . . . face-recognition algorithms . . . : For a glimpse of the enabling technology, see Clive Norris, Jade Moran, and Gary Armstrong, "Algorithmic Surveillance: The Future of Automated Visual Surveillance," in *Surveillance, Closed Circuit Television, and Social Control*, ed. Norris, Moran, and Armstrong, 255–75; and P. Remagnino and others, "Automatic Visual Surveillance of Vehicles and People," in *Advanced Video-Based Surveillance Systems*, 95–105. For more detailed information on performance, see Michael Negin and others, "An Iris Biometric System for Public and Personal Use," *Computer* 33 (February 2000): 70–75; Alex Pentland and Tanzeem Choudhury, "Face Recognition for Smart Environments," *Computer* 33 (February 2000): 50–55; P. Jonathon Phillips and others, "An Introduction to Evaluating Biometric Systems," *Computer* 33 (February 2000): 56–63; and Rahul Sukthankar and Robert Stockton, "Argus: The Digital Doorman," *IEEE Intelligent Systems* 16 (March/April 2001): 14–19.

116 The landmark case is . . . : *Katz v. United States*, 389 U.S. 347 (1967). A right to privacy exists in a situation when a citizen expects privacy and society at large considers that expectation reasonable.

116 If you think people who complain . . . : Erin Texeira, "Man Killed by Stray Bullet on New Year's," *Los Angeles Times*, January 2, 2001.

116 At midnight . . . : "Reveler's Gunfire Likely Cause of Girl's Death," *Houston Chronicle*, January 3, 2001.

116 Their patent application . . . : The U.S. Patent and Trademark Office awarded patent no. 5,973,998 ("Automatic Real-Time Gunshot Locator and Display System") to Robert L. Showen and Jason W. Dunham on October 26, 1999.

117 The process pinpoints gunshots . . . : Police in Redwood City, California, tested ShotSpotter by firing blank rounds at known locations throughout

a test area. According to police captain Jim Granucci, 80 percent of all the test shots were located within 15 yards of the known locations. Marshall Wilson, "Redwood City Gunshot Locator Passes Tests—Gets Trial Run," *San Francisco Chronicle,* July 11, 1996.

118 "What would you do if an officer . . . ": Trilon Technology, "Frequently Asked Questions," *ShotSpotter: The 9-1-1 Gunfire Alert System,* http://www.shotspotter.com/g-faq.html. The quotation is part of the answer to the question "Is ShotSpotter intended to prevent gunfire or to assist police in arresting those who shoot their firearms?"

118 . . . a ten-week trial in 1995: For details, see Marshall Wilson, "Redwood City Endorses Gunshot Locator System," *San Francisco Chronicle,* March 18, 1997; and Wilson, "Redwood City Gunshot Locator Passes Tests."

118 When a gunshot is detected . . . : Bobby Cuza, "Gadgets on Patrol against Crime," *Los Angeles Times,* June 9, 2000; and Willoughby Mariano, "Way to Locate Sources of Gunfire Shown," *Los Angeles Times,* December 30, 1999.

119 "the year's most innovative . . . ": "Using Technology in the Fight against Random Gunfire," *Microsoft PressPass,* April 3, 2000, http://www.microsoft.com/presspass/features/2000/04-03shotspotter.asp.

119 Although computers expedite . . . : For a concise discussion of early maps of crime data, see Phillips, "Geography of Crime;" and Robinson, *Early Thematic Mapping,* 156–70. Both authors examine maps of area data, but neither mentions "pin maps." For an overview of research and educational issues in contemporary GIS-based crime analysis, see Arthur Getis and others, "Geographic Information Science and Crime Analysis," *URISA Journal* 12 (spring 2000): 7–14.

119 The International Association of Chiefs of Police . . . : Michael D. Maltz, "From Poisson to the Present: Applying Operations Research to Problems of Crime and Justice," *Journal of Quantitative Criminology* 12 (1996): 3–61.

120 Perhaps the greatest impetus . . . : Robert K. Bratt, Joseph R. Lake, Jr., and Theresa Whistler, "Implementation of GIS at Local Law Enforcement Agencies," *Proceedings of the 1995 Arc/Info Users Conference,* http://www.esri.com/library/userconf/proc95/to200/p185.html.

120 As part of its outreach . . . : Harries, *Mapping Crime.* Distribution data from Jolene Hernon, National Institute of Justice, e-mail communication with the author, March 23, 2001.

121 The Chicago Police Department used . . . : Keith Harries warns that hot-spot ellipses should always be compared with the underlying point pattern. He also warns that hot-spot mapping, which focuses largely on street crime, might distract from white-collar crime. See Harries, *Mapping Crime,* 113–18.

121 For further information on ComStat, see Harries, *Mapping Crime,* 78–80; and Philip G. McGuire, "The New York Police Department ComStat

Process," in *Analyzing Crime Patterns: Frontiers of Practice,* ed. Victor Goldsmith and others (Thousand Oaks, Calif.: Sage Publications, 2000), 11–22.

122 ... he gave an insightful answer ... : Raymond Dussault, "Jack Maple: Betting on Intelligence," *Government Technology* 12 (April 1999): 26–28; quotation on 27. Also see Jack Maple and Chris Mitchell, *The Crime Fighter: Putting the Bad Guys Out of Business* (New York: Doubleday, 1999).

124 ... complaints against the police: "Compensating Abner Louima," *New York Times,* March 24, 2001. Louima, a Haitian immigrant, was the victim of a vicious attack by a police officer who shoved a broom handle up his rectum. Respect for police plummeted, several officers went to prison, and the city and the police union agreed to pay over $8 million in damages to settle Louima's lawsuit, thereby avoiding another embarrassing trial.

Chapter 8. Keeping Track

General Sources

Key sources for legal and administrative information on Megan's Law and sex-offender databases are Paul Koenig, "Does Congress Abuse Its Spending Clause Power by Attaching Conditions on the Receipt of Federal Law Enforcement Funds to a State's Compliance with 'Megan's Law'?" *Journal of Criminal Law and Criminology* 88 (1998): 721–65; and Bonnie Steinbock, "A Policy Perspective: Megan's Law—Community Notification of the Release of Sex Offenders," *Criminal Justice Ethics* 14 (summer-fall 1995): 4–9.

Critiques of the constitutionality and effectiveness of Megan's Laws include James Bickley and Anthony R. Beech, "Classifying Child Abusers: Its Relevance to Theory and Clinical Practice," *International Journal of Offender Therapy and Comparative Criminology* 45 (2001): 51–69; Alexander D. Brooks, "The Legal Issues: Megan's Law—Community Notification of the Release of Sex Offenders," *Criminal Justice Ethics* 14 (summer-fall 1995): 12–16; Charles J. Dlabik, "Convicted Sex Offenders: Where Do You Live? Are We Entitled to Know? A Year's Retrospective of Ex Post Facto Challenges to Sex Offender Community Notification Laws," *Nova Law Review* 22 (1998): 585–644; R. Karl Hanson, "What Do We Know about Sex Offender Risk Assessment?" *Psychology, Public Policy, and Law* 4 (1998): 50–72; and Philip H. Witt and others, "Sex Offender Risk Assessment and the Law," *Journal of Psychiatry and Law* 24 (1996): 343–77.

Sources of information on satellite tracking of offenders include Joseph Hoshen, Jim Sennott, and Max Winkler, "Keeping Tabs on Criminals," *IEEE Spectrum* 32 (February 1995): 26–32; Linda Johansson, "Invisible Chains," *UNESCO Courier* 51 (June 1998): 13–14; Dee Reid, "High-tech House Arrest: Electronic Ankle Bracelets Used to Monitor Prisoners under Home Detention," *Technology Review* 89 (July 1986): 12–14; and Robert E. Sullivan, Jr., "Reach Out and Guard

Someone: Using Phones and Bracelets to Reduce Prison Overcrowding," *Rolling Stone*, November 29, 1990, 51.

Notes

126   Within a week fifteen hundred people . . . : James Barron, "Vigil for Slain Girl, 7, Backs a Law on Offenders," *New York Times*, August 3, 1994. Actually a collection of laws regulating sex offenders, New Jersey's Megan's Law included a community notification statute based on a 1990 Washington State law. For an overview, see Joseph F. Sullivan, "Whitman Approves Stringent Restrictions on Sex Criminals," *New York Times*, November 1, 1994.

126   And those in Tier Three . . . : The law required the attorney general to draw up notification guidelines, which were published in late December; see Michael Booth, "Who Must Register and Who Should Know?" *New Jersey Law Journal*, December 26, 1994, 16.

127   "release relevant information . . . ": "Megan's Law," *Congressional Record* 142 (daily ed.; May 7, 1996): H4451–52; quotation on H4451.

127   For example, a May 1999 survey . . . : Devon B. Adams, "Summary of State Sex Offender Registry Dissemination Procedures," *Bureau of Justice Statistics Fact Sheet* publication no. NCJ 177620 (August 1999).

127   A June 1999 survey . . . : Jane A. Small, "Who Are the People in Your Neighborhood? Due Process, Public Protection, and Sex Offender Notification Laws," *New York University Law Review* 74 (1999): 1451–93.

127   By contrast, a May 2001 visit . . . : National Consortium for Justice Information and Statistics, "State Sex Offender Registry Websites," http://www.search.org/policy/nsor/state_webs.asp.

128   "does not maintain . . . ": "In Brief: The District," *Washington Post*, March 11, 2001.

128   In his April 2001 ruling . . . : Connecticut's attorney general asserted his intent to appeal the ruling, and as of May 8, 2001, the state's online sex-offender registry (www.state.ct.us/dps/Sor.htm) was still operating without any apparent recognition of risk level. But under increased court pressure, the state took its registry offline on May 18; see Paul Zielbauer, "Hartford's Sex-Offender Registry Shut Down after Judge's Order," *New York Times*, May 19, 2001. Meanwhile, state officials were deliberating a change in policy; see Stacey Stowe, "Talks Set to Begin on Sex Offender Site," *New York Times*, April 8, 2001, Connecticut Weekly section; and "Mend Megan's Law" [editorial], *Hartford Courant*, April 18, 2001.

128   "updates this information . . . ": Quotations in this and the following paragraph are from New York State Division of Criminal Justice Services, "Sex Offender Registry," http://criminaljustice.state.ny.us/nsor/index.htm, which also includes an explanation of New York's Sex Offender Registration Act. For information on the court ruling, see David W. Chen, "Federal

Judge Bars New York's Method of Classifying Sex Offenders," *New York Times*, May 8, 1998.

130 "naming and shaming": For an example, see "'Website Plan to Name Sex Offenders," *Herald* (Glasgow), November 15, 2000.

131 A prospective viewer . . . : See Office of the Attorney General, State of California Department of Justice, "Registered Sex Offenders (Megan's Law)," http://caag.state.ca.us/megan/index.htm.

131 For a description of the Fairfield, California, maps and restrictions on their use, see City of Fairfield, Police Department Sex Offender Database, http://www.ci.fairfield.ca.us/police/disclaimer.asp; and the Fairfield Police Department News Media Release "Megan's Law/Sex Registrant Maps on Police Website," December 11, 2000, http://www.ci.fairfield.ca.us/ announcements/files/PressRelease/PoliceDepartment/381568636/ PR0000.htm. For the map in figure 8.1, see specifically http://www.ci .fairfield.ca.us/police/map.asp?school_id=5.

132 More troublesome is the need . . . : For an overview of legal and ethical issues surrounding community notification, see Scott Matson, "Community Notification and Education," Center for Sex Offender Management, April 2001, http://www.csom.org/pubs/notedu.html.

132 "guarded villages": Etzioni suggests allowing offenders to live comparatively normal lives, hold steady jobs, and share social responsibilities within the guarded perimeter and "protective custody" of a community offering freedom from intrusion and shaming as long as they did not leave. See Amitai Etzioni, *The Limits of Privacy* (New York: Basic Books, 1999), 73–74.

132 For discussion of satellite tracking as a remedy for domestic violence as well as an alternative to parole for sex offenders, see David R. Kazak, "Home Monitoring Doesn't Stop Crime in Homes," *Chicago Daily Herald*, February 5, 2001; and Lori Montgomery and Daniel LeDuc, "Killing of Frederick Boy Stirs Debate about Freeing Molesters," *Washington Post*, January 21, 2001.

133 Satellite tracking is a significant advance . . . : Perhaps the earliest proposal for a location-tracking system required a network of pole-mounted receivers similar to the ShotSpotter system discussed in chapter 7; see Joseph A. Meyer, "Crime Deterrent Transponder System," *IEEE Transactions on Aerospace and Electronic Systems* 7 (1971): 2–22.

133 More advanced systems . . . : George Lane, "State to Test Satellite-based Tracking of Parolees," *Denver Post*, August 28, 1999. For discussion of distance restrictions, see Bill McGarigle, "The Walls Have Come Down," *Government Technology*, May 1997, http://www.govtech.net/publications/ gt/1997/may/may1997-geoinfo/may1997-geoinfo.phtml.

133 Compared to incarceration . . . : Tracking systems are usually leased, at a daily cost per detainee ranging from $7.50 to $18.00, according to various

sources; see Lane, "State to Test Satellite-based Tracking"; McGarigle, "The Walls Have Come Down"; Montgomery and LeDuc, "Killing of Frederick Boy Stirs Debate"; and Graham Rayman, "Monitoring Domestic Violence," *New York Newsday*, July 23, 1999, Queens edition.

134 "high-tech ball and chain": Heather Hayes, "The Long Arm of the Law," *FCW.com (Federal Computer Week)*, December 6, 1999, http://www.fcw .com/civic/articles/1999/CIVIC_120699_43.asp. Information about SMART is from Pro Tech's Web site (www.ptm.com).

134 Zoning can also restrict . . . : McGarigle, "The Walls Have Come Down." Building interiors are occasionally dead zones for wireless telephony; see Robert K. Morrow, Jr., and Theodore S. Rappaport, "Getting In," *Wireless Review* 17 (March 1, 2000): 42–44.

135 "Third-generation" systems . . . : Max Winkler, "Walking Prisons: The Developing Technology of Electronic Controls," *The Futurist* 27 (July/August 1993): 34–36.

135 For a description of the "Autoinjector," "Poison Vial," and similarly sinister implants, see Michael LaBossiere, "New Cyber Equipment for 2300 ad and Cyberpunk," *One Man's Views of 2300 ad,* http://www.crosswinds .net/~anch_stevec/newcyber.htm. For more contemporary approaches, see Hoshen, Sennott, and Winkler, "Keeping Tabs on Criminals"; and Ed Grabowski, "Electronic Monitoring of Prisoners" (November 1996), Computers and Law, University of Buffalo School of Law, http://wings .buffalo.edu/law/Complaw/CompLawPapers/grabowsk.html (site discontinued).

135 "invisible electronic fence": Canine systems typically use an electric shock to reinforce an audible warning; see Pati Simon Gelfman, "Invisible Fence for Dogs," *Family Handiman* 38 (January 1988): 44–45.

136 . . . to help railroads avoid rear-end collisions: Christine White, "On-Road, On-Time, and On-Line," *Byte* 20 (April 1995): 60–66; and Tom Sullivan, "PTC: Is FRA Pushing Too Hard?" *Railway Age* 200 (August 1999): 49–57.

136 . . . let an insurer monitor their driving: William Siuru, "OnStar to the Rescue," *Electronics Now* 69 (September 1998): 61–62. In 1998 the Progressive Insurance Company started to explore a role for GPS-based monitoring in setting auto insurance rates; see Greg Hassell, "Cheap Insurance Comes at a Price," *Houston Chronicle*, November 3, 1999, business section.

136 "apparatus for tracking . . . ": The U.S. Patent and Trademark Office issued patent no. 5,629,678 ("Personal Tracking and Recovery System") to Paul A. Gargano and others on May 13, 1997. The quotation is from the abstract of the patent. Also see Mark Harrington, "Alliance Boosts Monitoring System," *Newsday*, September 20, 2000; and Chris Trumble, "GPS Tracking Is Only Skin Deep," *Smart Computing* 11 (April 2000): 6.

136   Civil libertarians promptly warned . . . : Kurt Kleiner, "They Can Find You: GPS Implants Will Make It Easy to Pinpoint People," *New Scientist* 165 (January 8, 2000): 7. Also see Richard Stenger, "Tiny Human-Borne Monitoring Device Sparks Privacy Fears," *CNN.com*, December 20, 1999, http://www.cnn.com/1999/TECH/ptech/12/20/implant.device/index.html.

136   "new locational e-slavery": Susan L. Cutter, "President's Column—Big Brother's New Handheld," *AAG Newsletter* 36 (May 2001): 3–4; quotation on 3.

136   What loving son or daughter . . . : The preorder price of $299 did not include tax and shipping; see Digital Angel, "Preorder and Reserve," http://www.digitalangel.net/contact/preorder.htm.

136   Offered in early 2001 . . . : Based on ADS press releases, a November 1, 2000, news story on *WorldNetDaily.com* described "a miniature sensor device designed to be implanted just under the skin" as well as the firm's plans to produce a "more sophisticated version . . . powered electromechanically through muscle movement." See "Digital Angel Unveiled: Human-Tracking Subdermal Implant Technology Makes Debut," archived at Direct Source Radio, http://www.directsourceradio.com/links/11012000/110120004.html.

136   "implantable triggerable transmitting device": Quotations and figure 8.3 are from the application for U.S. Patent no. 5,629,678, granted May 13, 1997.

137   "branding and stalking": Quotations are from Jerome E. Dobson, "What Are the Ethical Limits of GIS?" *GeoWorld* 13 (May 2000): 24–25.

138   In 1996 the FCC . . . : Under the original mandate, carriers had to be able to estimate location to within 125 meters for 67 percent of callers by October 2001, and for all callers by the end of 2002. See Dee Ann Divis, "Privacy Matters: Data, Mobile Commerce, GIS," *Geospatial Solutions* 10 (October 2000): 18–20. Extended 911 wireless regulations are more complex than most writers suggest, and the FCC has revised them several times since 1996. When carriers failed to meet an October 2001 deadline, the FCC granted a generous extension but imposed various intermediate milestones intended to encourage nearly complete compliance by the end of 2005. See Suzanne King and David Hayes, "Deadline Extended for 911 Technology," *Kansas City Star,* October 9, 2001; and "FCC Acts on Wireless Carrier and Public Safety Requests Regarding Enhanced Wireless 911 Services." *FCC News,* October 5, 2001, http://www.fcc.gov/Bureaus/Wireless/News_Releases/2001/ nrwl0127.html.

138   The original order let . . . : Kevin McLaughlin, "Wireless Carriers Announce Location Tech Plans," *Business 2.0,* November 10, 2000, http://www.business2.com/content/channels/technology/2/11/10/22478.htm.

138 For examples of location-based services, see Jay Benson, "LBS Technology Delivers Information Where and When It's Needed," *Business Geographics* 9 (February 2001): 20–22; and Jonathan W. Lowe, "The Power of Babble: Congregating around LBS," *Geospatial Solutions* 11 (February 2001): 46–51.

139 Loss of privacy is inevitable . . . : For hypothetical examples of privacy issues raised by wireless tracking based on GPS, see Mike France and Dennis K. Berman, "Big Brother Calling: Location Technology in Devices Such as Cell Phones," *Business Week* no. 3700 (September 25, 2000): 92–98; and Robert Poe, "Location Disorder," *Business 2.0*, March 26, 2001, http://www.business2.com/technology/2001/03/28411.htm. For wider insights on the effects of GPS and mobile telephony on personal privacy, see Roger Clarke, "Person-Location and Person-Tracking: Technologies, Risks, and Policy Implications," *Roger Clarke's Dataveillance and Information Privacy Pages* (hosted by Australian National University), http://www.anu.edu.au/people/Roger.Clarke/DV/PLT.html.

Chapter 9. Addresses, Geocoding, and Dataveillance

General Sources

Sources on the use and social implications of clustering include Jon Goss, " 'We Know Who You Are and We Know Where You Live': The Instrumental Rationality of Geodemographic Systems," *Economic Geography* 71 (1995): 171–98; Erik Larson, *The Naked Consumer: How Our Private Lives Become Public Commodities* (New York: Henry Holt and Company, 1992); Michael J. Weiss, *The Clustered World: How We Live, What We Buy, and What It All Means about Who We Are* (Boston: Little, Brown, 2000); and Michael J. Weiss, *The Clustering of America* (New York: Harper and Row, 1988).

Key sources on data privacy include Roger A. Clarke, "Information Technology and Dataveillance," *Communications of the Association for Computing Machinery* 31 (1988): 498–512; Michael R. Curry, "The Digital Individual and the Private Realm," *Annals of the Association of American Geographers* 87 (1997): 681–99; Jon Goss, "Marketing the New Marketing," in *Ground Truth: The Social Implications of Geographic Information Systems*, ed. John Pickles (New York: Guilford Press, 1995), 130–70; Gary T. Marx, *Undercover: Police Surveillance in America* (Berkeley: University of California Press, 1988); Harlan J. Onsrud, Jeff P. Johnson, and Xavier R. Lopez, "Protecting Personal Privacy in Using Geographic Information Systems," *Photogrammetric Engineering and Remote Sensing* 60 (1994): 1083–95; and Harlan J. Onsrud, "The Tragedy of the Information Commons," in *Policy Issues in Modern Cartography*, ed. D. R. Fraser Taylor (New York: Pergamon, 1998), 141–58.

Notes

141   According to recent estimates . . . : Tabulations are from ESRI Business
       Information Solutions, "Free Zip Code Report" (based on 2000 projec-
       tions),http://infods.com/freedata. In early 2002, ESRI Business Infor-
       mation Solutions acquired the marketing services division of CACI,
       which had developed ACORN.

142   Two separate previews actually . . . : Cluster descriptions and quotations
       are from ClaritasExpress, "You Are Where You Live," http://www.cluster2
       .claritas.com/YAWYL.

142   . . . reports only the top five . . . : A list might include fewer than five clus-
       ters because the Web site does not list lifestyle types that characterize less
       than 5 percent of an area's population.

142   Although "Winner's Circle" might connote . . . : Advertising consultant
       Robin Page crafted names for PRIZM's clusters; see Philip H. Dougherty,
       "ZIP Area: Key to Markets," *New York Times*, July 17, 1980.

144   . . . named ACORN . . . : Clusters and quotations from CACI, "IDS On
       the Web," http://www.infods.com.

145   Each piece of mail . . . : For insights on the advantages of custom-screened
       mailing lists, see Christopher Carey, "More and More Companies Are
       Reaching Their Customers through Direct Mail," *St. Louis Post-Dispatch*,
       December 7, 1998; and Amy Merrick, "New Population Data Will Help
       Marketers Pitch Their Products," *Wall Street Journal*, February 14, 2001.
       According to Merrick, Hyundai pays $200,000 annually for customized
       mailing lists that maximize its response rate for test-drive promotions.

145   Political consultants are equally eager . . . : For discussion of targeted
       mailing in political campaigns, see David Beiler, "Precision Politics,"
       Campaigns & Elections 10 (February/March 1990): 33–36, 38; Ron Fau-
       cheux, "Hitting the Bull's Eye," *Campaigns & Elections* 20 (July 1999): 20–
       25; and Leslie Wayne, "Voter Profiles Selling Briskly as Privacy Issues Are
       Raised," *New York Times*, September 9, 2000.

146   "Harvard-educated computer whiz": Weiss, *Clustering of America*, 10.

146   "accounted for 87 percent . . . ": Ibid., 11. For a concise introduction to the
       use of factor analysis in detecting and describing lifestyle clusters, see
       David J. Curry, *The New Marketing Research Systems: How to Use Strategic
       Database Information for Better Marketing Decisions* (New York: John Wiley
       and Sons, 1993), 203–6.

146   PRIZM's success is apparent . . . : National Decision Systems introduced
       MicroVision in 1990. VNU, the Dutch company that acquired Claritas in
       1986 and National Decision Systems in 1997, merged NDS into Claritas
       in 1999. For further information on Claritas, see "Company Info—Clari-
       tas History," http://www.claritas.com/index.html. For a discussion of
       clustering outside North America, see Weiss, *Clustered World*.

147 For additional information on the U.S. Postal Service's marketing of National Change of Address (NCOA) data to private-sector mailing-list companies, see "Move Update: Keeping Up with Your Moving Customers," U.S. Postal Service booklet (August 2000); and "Panel: USPS Violates Privacy and Law by Making Public Change-of-Address Orders," *Direct Marketing* 55 (February 1993): 10–11.

147 "'black box' mechanics . . . ": Goss, "'We Know Who You Are," 187.

147 Defined more broadly . . . : Definitions and populations are from U.S. Census Bureau, "Geographic Terms and Concepts," http://www.census.gov/geo/www/tiger/glossry2.html.

147 Blocks aggregate conveniently . . . : The "optimum" populations of block groups and census tracts are fifteen hundred and four thousand, respectively. Ibid.

148 This assumption is crucial . . . : For a concise overview of the inverse relationship between the geographic precision of small-area data and the demographic precision of detailed cross-tabulations, see John Kavaliunas, "Get Ready to Use Census 2000 Data," *Marketing Research* 12 (fall 2000): 42–43.

148 A computer calculates geographic . . . : See, for example, ESRI BIS, "Demographic Update Methodology," http://www.infods.com/methodology/.

149 Although the agency never discloses . . . : As part of the "Fifth Count" round of tabulations for the 1970 census, the Census Bureau released estimated counts based on the long-form questionnaires filled out by 5-, 15-, and 20-percent samples of households. Estimates focused on race, age, sex, marital status, income, education, and condition of housing and were distributed on magnetic tape. Fifth Count File 5A contained estimates for three-digit ZIP Codes for the entire country, and Fifth Count File 5B contained estimates for approximately 12,500 five-digit ZIP codes within metropolitan areas. See National Archives and Records Administration, Center for Electronic Records, Reference Report no. 11, http://www.nara.gov/nara/electronic/cen1970.html.

149 "the sweetheart deal of the century": Edward Spar, president of Market Statistics, quoted in Larson, *Naked Consumer,* 45.

151 The Internal Revenue Service's SOI Products & Services directory is available online: http://www.irs.gov/tax_stats/.

151 "largest collection of U.S. consumer . . . ": "Selling Is Getting Personal," *Consumer Reports* 65 (November 2000): 16–20; quotation on 18. Also see Robert O'Harrow, Jr., "Eye at the Keyhole: Privacy in the Digital Age," *Washington Post,* March 8, 1998.

151 "new resident direct marketing": Quotations are from Moving Targets, "How We Target Mailings to the Latest and Best New Movers," http://www.movingtargets.com/moreinfo2.html.

151 . . . Claritas clients find . . . : Susan Mitchell, "Birds of a Feather," *Ameri-*

can Demographics 17 (February 1995): 40–48; and Tom Spencer and David Tedrow, "Capture Customers for Life," *Business Geographics* 8 (September 2000): 28–30.

151   . . . an experienced snoop can dig up dirt . . . : Marx, *Undercover*, 210–11.

151   Home telephone numbers . . . : Robert O'Harrow, Jr., "A Hidden Toll on Free Calls: Lost Privacy," *Washington Post,* December 19, 1999; and Gary Angel and Joel Hadary, "Using Card Transaction Data," *American Demographics* 20 (August 1998): 38–41.

151   For an overview of address-matching technology and applications, see William J. Drummond, "Address Matching: GIS Technology for Mapping Human Activity Patterns," *Journal of the American Planning Association* 61 (1995): 240–51.

152   Web browsing is also under surveillance . . . : For examples, see Amy Harmon, "Software to Track E-Mail Raises Privacy Concerns," *New York Times,* November 22, 2000.

152   "Geotargeting": Stefanie Olsen, "Yahoo Ads Close in on Visitors' Locale," *CNET News.com,* June 27, 2001, http://news.com.com/2100-1023-269155 .html.

152   For a concise explanation of cookies, see Glenn Fleishman, "Fresh from Your Browser's Oven," *New York Times,* July 15, 1999.

152   . . . customize the banner ads . . . : Bob Tedeschi, "Critics Press Legal Assault on Tracking of Web Users," *New York Times,* February 7, 2000.

152   Largely benign, cookies can reveal . . . : Lance Gay, "Drug Czar Asks Congress to Reopen the 'Cookie' Jar," *Syracuse Post-Standard,* July 16, 2000.

152   "cache attack": Ian Austen, "Study Finds That Caching by Browsers Creates a Threat to Surfers' Privacy," *New York Times,* December 14, 2000.

152   "systematic use of personal data . . . ": Clarke, "Information Technology and Dataveillance," 499.

152   In a capitalist milieu . . . : Goss, "We Know Who You Are," 192–93.

153   . . . critic of GIS-based behavioral modeling . . . : For example, Stephen Graham, "Surveillant Simulation and the City: GIS and Urban Panopticism" (paper presented at the National Center for Geographic Information and Analysis Conference on Spatial Technologies, Geographic Information, and the City, Baltimore, Md., September 9–11, 1996, http://www.ncgia.ucsb.edu/conf/BALTIMORE/authors/graham/paper .html).

153   . . . widespread use of integrated systems . . . : See Curry, "The Digital Individual and the Private Realm."

153   For examples of clusters that reinforce an address's prestige, see Beth Daley, "Residents Ride Latest Wave of New Numbers," *Boston Globe,* February 2, 1998; and Emily Wax, "Mail, the Great Equalizer," *New York Times,* July 12, 1998, Queens edition.

153   . . . integrating diverse databases fosters . . . : See Onsrud, "Information Commons."

153   "what is agreed to be 'smart business practices' . . . ": Harlan J. Onsrud, "Ethical Issues in the Use and Development of GIS," *Proceedings of GIS/LIS '97*, 400–401; quotation on 401.

Chapter 10. Case Clusters and Terrorist Threats

General Sources

Sources addressing environmental "right-to-know" laws include Bradley C. Karkkainen, "Information as Environmental Regulation: TRI and Performance Benchmarking, Precursor to a New Paradigm," *Georgetown Law Journal* 89 (2001): 257–370; and Sidney M. Wolf, "Fear and Loathing about the Public Right to Know: The Surprising Success of the Emergency Planning and Community Right-to-Know Act," *Journal of Land Use and Environmental Law* 11 (1996): 217–324.

Basic works on disease mapping are Andrew B. Lawson and Fiona L. R. Williams, *An Introductory Guide to Disease Mapping* (New York: John Wiley, 2001); and Steven M. Teutsch and R. Elliott Churchill, eds., *Principles and Practice of Public Health Surveillance* (New York: Oxford University Press, 2000).

Useful references on the use of GIS in public health include Keith C. Clarke, Sara L. McLafferty, and Barbara J. Tempalski, "On Epidemiology and Geographic Information Systems: A Review and Discussion of Future Directions," *Emerging Infectious Diseases* 2 (1996): 85–92; P. E. R. Dale and others, "An Overview of Remote Sensing and GIS for Surveillance of Mosquito Vector Habitats and Risk Assessment," *Journal of Vector Ecology* 23 (1998): 54–61; Andrew Lovett and others, "Improving Health Needs Assessment Using Patient Register Information in a GIS," in *GIS and Health*, ed. Anthony G. Gatrell and Markku Löytönen (London: Taylor and Francis, 1998), 191–203; Thomas J. McGinn, 3rd, Peter Cowen, and David W. Wray, "Geographic Information Systems for Animal Health Management and Disease Control," *Journal of the American Veterinary Medicine Association* 209 (1996): 1917–21; Thomas B. Richards, Charles M. Croner, and Lloyd F. Novick, "Atlas of State and Local Geographic Information Systems (GIS) Maps to Improve Community Health," *Journal of Public Health Management and Practice* 5 (March 1999): 2–8; Gerard Rushton, Gregory Elmes, and Robert McMaster, "Considerations for Improving Geographic Information System Research in Public Health," *URISA Journal* 12 (spring 2000): 31–49; Robert K. Washino and Byron L. Wood, "Application of Remote Sensing to Arthropod Vector Surveillance and Control," *American Journal of Tropical Medicine and Hygiene* 50, supplement (1994): 134–44; and Paul Wilkinson and others, "GIS in Public Health," in *GIS and Health*, ed. Anthony G. Gatrell and Markku Löytönen (London: Taylor and Francis, 1998), 179–89.

Notes

155 . . . Snow made his famous dot map . . . : Howard Brody and others, "Map-Making and Myth-Making in Broad Street: The London Cholera Epidemic, 1954," *Lancet* 356 (2000): 64–68. Also see John Snow, *On the Mode of Communication of Cholera*, 2nd ed. (London: Churchill, 1855).

155 Even so, his pin map . . . : For an example, see Laura Lang, *GIS for Health Organizations* (Redlands, Calif.: ESRI Press, 2000), 13–15. In addition to a picture of Snow and a facsimile of his original map, Lang includes a map of Snow's 1854 data as plotted with a contemporary GIS. For a much earlier disease map, see Frank A. Barrett, "Finke's 1792 Map of Human Diseases: The First World Disease Map?" *Social Science and Medicine* 50 (2000): 915–21.

155 Moreover, contemporary investigators . . . : Kari S. McLeod, "Our Sense of Snow: The Myth of John Snow in Medical Geography," *Social Science and Medicine* 50 (2000): 923–35.

156 "geographic patterns of cancer . . . ": Thomas J. Mason and others, *Atlas of Cancer Mortality for U.S. Counties, 1950–1969*, Department of Health, Education and Welfare publication no. (NIH) 75–780 (Washington, D.C., 1975), v.

156 "perhaps the greatest value of the maps . . . ": Ibid.

156 "previously unnoticed clusters . . . ": Linda Williams Pickle and others, *Atlas of United States Mortality*, Department of Health and Human Services publication no. (PHS) 97–1015 (Hyattsville, Md., 1996), 1.

156 the "field studies" cited . . . : William J. Blot and others, "Lung Cancer after Employment in Shipyards during World War II," *New England Journal of Medicine* 299 (1978): 620–24; and Deborah M. Winn and others, "Snuff Dipping and Oral Cancer among Women in the Southern United States," *New England Journal of Medicine* 304 (1981): 745–49.

156 "generate[ed] etiological hypotheses": Pickle and others, *Atlas of United States Mortality*, 1.

156 Sources of information on the Onondaga County Breast Cancer Mapping Project include Sue Weibezahl, "Cancer Survey Will Be Largest in Nation," *Syracuse Herald Journal*, February 19, 1999; and Nicholas J. Pirro, Memorandum to members of the Onondaga County Legislature Onondaga County, New York, Press Releases, September 15, 1999, http://www3.co.onondaga.ny.us/Press/press.releases/budmess.html.

156 Although our county executive . . . : Nicholas J. Pirro, "County Executive Pirro Delivers His 2000 State of the County Message," Onondaga County, New York, Press Releases, March 6, 2000, http://www3.co.onondaga.ny.us/Press/press.releases/2000306.html.

157 . . . ZIP code-level breast cancer map . . . : See Richard Perez-Pina, "Breast Cancer Is Pinpointed by ZIP Code," *New York Times*, April 12, 2000; and New York State Department of Health, New York State Cancer Surveil-

lance Improvement Initiative, http://www.health.state.ny.us/nysdoh/
cancer/csii/main.htm.

157　...a disease especially common...: Martin Kulldorff and others, "Breast Cancer Clusters in the Northeast United States: A Geographic Analysis," *American Journal of Epidemiology* 146 (1997): 161–70.

157　"don't really identify cancer clusters..." : Quoted in Perez-Pina, "Breast Cancer Is Pinpointed."

157　"it's a sin to have..." and "it's a first step..." : Quoted in Dan Fagin, "Reading the Maps: State Tracks Breast Cancer but Warns Causes Still Elusive," *New York Newsday*, April 12, 2000.

157　While relatively detailed maps...: See, for example, Steven D. Stellman and others, "Breast Cancer Risk in Relation to Adipose Concentrations of Organochlorine Pesticides and Polychlorinated Biphenyls in Long Island, New York," *Cancer Epidemiology, Biomarkers, and Prevention* 9 (2000): 1241–50. For reactions of breast cancer activists to the Stellman study, see John Rather, "Breast Cancer Groups Question New Study," *New York Times*, December 3, 2000, Long Island edition.

158　"although the finding of no spatial clustering..." : Linda M. Timander and Sara McLafferty, "Breast Cancer in West Islip, NY: A Spatial Clustering Analysis with Covariates," *Social Science and Medicine* 46 (1998): 1623–35; quotation on 1634.

159　A "map hacker" could...: Marc P. Armstrong and Amy J. Ruggles, "Map Hacking: On the Use of Inverse Address-Matching to Discover Individual Identities from Point-Mapped Information Sources" (paper presented at the Geographic Information and Society Conference, University of Minnesota—Twin Cities, June 20–22, 1999, sponsored by the National Center for Geographic Information and Analysis), http://www.socsci.umn.edu/~bongman/gisoc99/new/armstrong.htm.

159　Researchers can protect privacy...: Marc P. Armstrong, Gerard Rushton, and Dale L. Zimmerman, "Geographically Masking Health Data to Preserve Confidentiality," *Statistics in Medicine* 18 (1999): 497–525.

159　For an overview of West Nile virus and its spread in the northeastern United States, see Martin Enserink, "The Enigma of West Nile," *Science* 290 (2000): 1482–84. Also see "Guidelines for the Surveillance, Prevention, and Control of West Nile Virus Infection—United States," *Morbidity and Mortality Weekly Report* 49 (January 21, 2000): 25–28.

160　Clusters of child pedestrian accidents...: Mary Braddock and others, "Using a Geographic Information System to Understand Child Pedestrian Injury," *American Journal of Public Health* 84 (1994): 1158–61.

160　The researchers validated the map...: Gregory E. Glass and others, "Environmental Risk Factors for Lyme Disease Identified with Geographic Information Systems," *American Journal of Public Health* 85 (1995): 944–48.

160　...a GIS-based model for high-risk areas...: Ellen K. Cromley and others, "Residential Setting as a Risk Factor for Lyme Disease in a Hyperendemic Region," *American Journal of Epidemiology* 147 (1998): 472–77.

161 "housing quality and maintenance practices": Daniel A. Griffith and others, "A Tale of Two Swaths: Urban Childhood Blood-Lead Levels across Syracuse, New York," *Annals of the Association of American Geographers* 88 (1998): 640–65; quotations on 661.

161 . . . "right-to-know" laws: Despite the legislated openness, for some data you might need to file a Freedom of Information Act (FOIA) request or join a volunteer fire department.

162 . . . a computer model of local aquifers: For examples, see Mark Monmonier, *Cartographies of Danger: Mapping Hazards in America* (Chicago: University of Chicago Press, 1997), 127–47.

162 For a concise overview of public participation GIS, see Nancy J. Obermeyer, "The Evolution of Public Participation GIS," *Cartography and Geographic Information Systems* 25 (1998): 65–66.

162 For an examination of the need for partnerships in PPGIS, see Trevor Harris and Daniel Weiner, "Empowerment, Marginalization, and 'Community-integrated' GIS," *Cartography and Geographic Information Systems* 25 (1998): 67–76.

162 Community Mapping Assistance Project: See the project's Web site (www.cmap.nypirg.org); and Melissa P. McNamara, "Project Gives Small Nonprofit Groups a Big-Time Mapping Tool," *New York Times,* July 20, 2000.

163 . . . thyroid cancer in eastern Washington: William D. Henriques and Robert F. Spengler, "Locations around the Hanford Nuclear Facility Where Average Milk Consumption by Children in 1945 Would Have Resulted in an Estimated Median Iodine-131 Dose to the Thyroid of 10 Rad or Higher, Washington," *Journal of Public Health Management and Practice* 5 (March 1999): 35–36. Also see Washington Department of Health Individual Dose Assessment, "Hanford IDA Questions and Answers" http://www.doh.wa.gov/ida/q&a.htm.

164 In North Dakota, for example . . . : Leona Kuntz and Alana Knudson-Buresh, "Adequate Prenatal Care Rates in North Dakota, 1991–1995," *Journal of Public Health Management and Practice* 5 (March 1999): 23–24.

164 For Hillsborough County, Florida . . . : Jason Devine, William K. Gallo, and Henry T. Janowski, "Identifying Predicted Immunization 'Pockets of Need,' Hillsborough County, Florida, 1996–1997," *Journal of Public Health Management and Practice* 5 (March 1999): 15–16.

164 In DeKalb County, Georgia . . . : Michael Y. Rogers, "Using Marketing Information to Focus Smoking Cessation Programs on Specific Census Block Groups along the Buford Highway Corridor, DeKalb County, Georgia, 1996," *Journal of Public Health Management and Practice* 5 (March 1999): 55–57.

164 . . . on a North Carolina military base . . . : Kelly T. McKee, Jr., "Application

of a Geographic Information System to the Tracking and Control of an Outbreak of Shigellosis," *Clinical Infectious Diseases* 31 (2000): 728–33.

165 And when an epizootic starts spreading rapidly . . . : For an example of a GIS designed to combat epidemics in animals, see Robert L. Sanson, Roger S. Morris, and Mark W. Stern, "EpiMAN-FMD: A Decision Support System for Managing Epidemics of Vesicular Disease," *Revu scientifique et technique* (International Office of Epizootics) 18 (1999): 593–605.

165 "The classic case was the burying . . . ": Quoted in Chris Partridge, "How Computers Can Help to Beat Foot-and-Mouth," *Times* (London), June 21, 2001.

166 For a concise overview of the Rajneeshee salmonella attack, see Steven M. Block, "The Growing Threat of Biological Weapons," *American Scientist* 89 (2001): 28–37. Also see Charles Marwick, "Scary Scenarios Spark Action at Bioterrorism Symposium," *Journal of the American Medical Association* 281 (1999): 1071–73.

167 "the poor man's atom bomb": Quoted in an Associated Press story published in New York's *Newsday*, August 15, 1998.

167 Also vulnerable is the nation's food supply . . . : Cheryl Pellerin, "The Next Target of Bioterrorism: Your Food," *Environmental Health Perspectives* 108 (2000): A126–29.

167 Average incubation periods for smallpox and anthrax are from Johns Hopkins Center for Civilian Biodefense Strategies agent fact sheets, http://www.hopkins-biodefense.org/pages/agents/agent.html.

167 But the two postal workers . . . : Johns Hopkins Bloomberg School of Public Health, "Researchers Examine Deaths of Two Postal Workers from Inhalational Anthrax," http://www.jhsph.edu/pubaffairs/press/postal_workers.htm.

167 . . . a long-term economic burden . . . : Arnold F. Kaufmann, Martin I. Meltzer, and George P. Schmid, "The Economic Impact of a Bioterrorist Attack: Are Prevention and Postattack Intervention Programs Justifiable?" *Emerging Infectious Diseases* 3 (April—June 1997): 83–94.

167 "whether the source is bioterrorism . . . ": Donald A. Henderson, "A New Strategy for Fighting Biological Terrorism," *The National Academies in Focus* 1 (spring 2001): 20–21.

Epilogue. Locational Privacy as a Basic Right

Notes

170 A suitably odious villain . . . : See Catherine Greenman, "The Car Snitched. He Sued," *New York Times*, June 28, 2001; and Paul Zielbauer, "Agency Protests Company's Fines on Speeders," *New York Times*, July 5, 2001.

170 . . . hired private firms like Lockheed-Martin . . . : Molly Ball, "Police Tout Success of Red-Light Cameras," *Washington Post*, August 9, 2001, Howard extra; and Dana Wilkie, "Red-Light Cameras Debated," *San Diego Union-Tribune*, August 1, 2001.

171 Image analysis software . . . : Andy Newman, "Those Dimples May Be Digits," *New York Times*, May 3, 2001. The first serious controversy in the United States arose in Tampa, Florida. See Dana Canedy, "Tampa Scans the Faces in Its Crowds for Criminals," *New York Times*, July 4, 2001; and Miki Meek, "You Can't Hide Those Lying Eyes in Tampa," *U.S. News and World Report* 131 (August 6, 2001): 20.

171 For an example of unintended consequences of red-light cameras, see the Protector antiphoto license plate cover offered by Jammers Store.com, http://www.jammersstore.com/anti_photo.htm.

172 "the right to be let alone": The phrase appears on the first page of a seminal article by Brandeis and a coauthor, who later repeat the wording inside quotation marks and attribute the last four words to a "Judge Cooley," cited as "*Cooley on Torts*, 2d ed., p. 29." See Samuel Warren and Louis D. Brandeis, "The Right to Privacy," *Harvard Law Review* 4 (1890): 193–220; quotation on 193 and 195.

172 For an assessment of the Gramm-Leach-Bliley Financial Services Modernization Act of 1999, see John Schwartz, "Privacy Policy Notices Are Called Too Common and Too Confusing," *New York Times*, May 7, 2001.

172 For discussion of cavalier disclosure of credit reports and medical records, see Simpson Garfinkle, *Database Nation: The Death of Privacy in the 21st Century* (Sebastopol, Cal.: O'Reilly, 2000), 21–29, 125–53.

172 "the next best thing": Erik Larson, *The Naked Consumer: How Our Private Lives Become Public Commodities* (New York: Henry Holt, 1992), 53–54; quotation on 53.

173 For a promising example of guidelines addressing maps and locational privacy, see Julie Wartell and J. Thomas McEwen, *Privacy in the Information Age: A Guide for Sharing Crime Maps and Spatial Data*, report no. NCJ-188739 (Washington, D.C.: National Institute of Justice, 2001).

Index